The Significance and Social Impact of Quarrying in Shropshire in the 19th and 20th Centuries

The Significance and Social Impact of Quarrying in Shropshire in the 19th and 20th Centuries

By

Dr. Robert S. Galloway

Cambridge
Scholars
Publishing

The Significance and Social Impact of Quarrying in Shropshire in the 19th and 20th Centuries

By Dr. Robert S. Galloway

This book first published 2019

Cambridge Scholars Publishing

Lady Stephenson Library, Newcastle upon Tyne, NE6 2PA, UK

British Library Cataloguing in Publication Data
A catalogue record for this book is available from the British Library

ISBN (10): 1-5275-3303-4
ISBN (13): 978-1-5275-3303-5

The book is dedicated to my Late mother Mrs Betty Galloway 1929-2015.

For her encouragement and financial assistance, without, which I could not have completed my PhD.

CONTENTS

LIST OF FIGURES

Chapter 1

Chapter 3

Chapter 4

Chapter 5

Chapter 6

List of Tables

Chapter 3

Chapter 4

Chapter 5

Chapter 6

Appendices:

ABSTRACT

This book investigates the technological innovations associated with the quarrying industry of Shropshire during the 19th and 20th centuries, from the extraction of rough-cut limestone, to greywacke used as road stone, sandstone flags found in vernacular buildings and finally to dimension stone. Examples of these minerals can be found in Shropshire, so such a breadth of geology has made this county unique. Knowledge of geology and minerals is integral and united in the quarrying industry. The inaccessibility and remote location of the raw materials has made the quarrying industry in Shropshire different from factory based industries.

The research found that legislation pertaining to the quarrying industry in Shropshire during the 19th century was non-existent, however, initial laws introduced in the 1890's attempted to structure the quarrying industry, its workforce and the local population. Research shows how this legislation would develop, assist and control the burgeoning quarrying industry, until today, planning applications direct industry to address care and consideration for the natural landscape, ecology, as well as saving important archaeological material today.

Evidence gained from original documents has enabled me to expand the knowledge base and broaden the picture, already begun by other researchers into varied aspects of Shropshire's industrial past and geology. Therefore, this book sets out to show the complete historical and archaeological importance of Shropshire's quarrying industry, the lives of ordinary people, land owners and entrepreneurs.

The quarrying industry, associated firstly with agriculture, rose to the ever increasing needs of the industrial revolution by moving from manpower to machinery driven by steam. Shropshire's industrial revolution has been widely researched, however, focus on the bulk mineral quarrying industry has not previously been undertaken. Quarrying and mining for minerals has never been studied before, although many resources can be found in archives, libraries and gained from personal accounts including from historical groups. All of which show the depth and importance of the quarrying industry economically both to Shropshire and the whole country. This book shows how the quarrying industry developed from individual families, to large entrepreneurial groups which swallowed-up these individually owned quarries. Minerals were initially extracted from the landscape by people with basic tools, until finally large steam, then petrol, driven machinery was developed. Over time it was necessary to change from horse and cart, to canal, to railway and eventually lorry transport to export the increasing quantities of extracted minerals, as a result of the financial and economic development produced by the entrepreneurial companies.

However, there are consequences that come from the quarrying industry such the effects on the ecology of the immediate and surrounding areas. Not only in terms of ecology, but the effects of former quarrying and mining in the 19th century, which can and as coursed subsidence's in particular parts of Shropshire, the latest being in the 1980s. This is why today legislation is very important, quarry companies today have to apply for planning permission and comply with planning regulations, but saying this the quarrying industry was and is still very important to Shropshire. One today has only to look at the excellent buildings and bridges in Shrewsbury constructed from Grinshill quarried stone from these it is clear to see the quality of the mineral.

ACKNOWLEDGEMENTS

I wish to acknowledge the debt I owe Dr Roger. H. White, my supervisor. To the Shropshire Archives, Staffordshire Record Office, The National Library of Wales. Ironbridge Gorge Library/Archives and The University of Birmingham Library.

Interviews and site visits with Mrs Joan Zorn, The Friends of Llanymynech Limeworks, Trevor Kirkham, Grinshill Quarries, Councillor Ted Clarke of Bayston Hill, Column and Sutton, Harriet Devlin, MBE and Grahame French of Shropshire County Council. I would also like to acknowledge the support of my dyslexic tutor Frances Godley.

CHAPTER 1

CONTEXT

1.1 Introduction

This book aims to provide an innovative analysis of the development of the quarrying industry of Shropshire within the 19th and 20th centuries. It will investigate all the technological revolutions associated with the extraction of limestone, sandstone, greywacke, and dolerite within the quarrying industry. Although my research has been into the 19th and 20th century quarrying industry, to give an overall account, one has to look further back in time when quarrying was at an earlier stage and transport was on the backs of people and pack-horses. The processes involved in quarrying for minerals are unlike those experienced in factory-based industries, because the raw material, in some cases, may be remotely located and is often very inaccessible.

1.2 Geology

Many scholars have written about the geology of Shropshire, such as the pioneers, Sir Roderick. I. Murchison, in the early 19th century, Professor Charles Lapworth in the early 20th century and recently Mr Peter Toghill, one of the foremost geologists of the 21st century. All these men and others have described the geological composition of Shropshire. Toghill describes the geological formation of sandstone at Grinshill, the limestone at Llanymynech Hill, Wenlock Edge and Lilleshall, greywacke at Bayston Hill and the dolerite at Brown Clee Hill Quarries. Toghill also talks about the individual colours, compositions, grains and the uses of the different minerals (Toghill, 2000).

1.3 Historical Background

A sound basis for this research was gained by investigating the archaeological and historical background of each quarry site, from monastic origins, agricultural, individual or family quarry owners, through the various landowners, until the last quarry industrialists. This provides a wider understanding of the background of the origins of mineral extraction in Shropshire. The research will establish how the industry worked and expanded through the two centuries.

The research established that the disparate workforce, men, women and children, was drawn from both outlying and local villages. From this, various questions arose as to: why was the workforce drawn from outlying villages and towns, and if they lived a distance away from the quarry or mining sites, how long did it take to get to work, how long was their working day and was it impractical to walk back home at the end of a shift? In this case where did they live for the working week, on site, or in local villages or towns?

Documentation from archives provided information about quarrying companies that built houses for their employees, and indicated how near, or far from the quarry, the accommodation was sited. Where the quarry owners provided accommodation for their employees, there were consequences for the workforce, which was that they had to work longer hours in the summer months in order to keep their homes. The quarry owners' reasoning was that hours were lost in winter months, when there was bad weather, which needed to be made up in the summer months during the longer daylight hours. If the workforce did not put in the extra summer working hours they and their families would be evicted from the company's housing. This also brought into question the number of hours and days worked during 19th century quarrying leading to the supposition that health and safety was not taken into consideration at that time.

Shropshire has been one of the most rural and sparsely populated counties in England, yet its geological composition offers a wealth of mineral resources. These are not only of local importance, but of national too, in particular greywacke, which is a limited resource in Britain, being found only in a few places in the country. The extraction of these different minerals has and continues to contribute to the local and national infrastructure of our society and the wealth of the county and country. However, minerals can only be worked where they are found, and their extraction, in populated areas causes conflict within society, either through the loss, or change to valued landscape, of habitats, or features of archaeological and historical importance. This is why Shropshire, because of its rural makeup, more so in the 19th century than today, was and still is ideal for mineral extraction. Shropshire has built new towns and villages and existing estates have expanded, but compared with other English counties, it is still very rural and less populated (Lapworth *et al,* 1910 and Toghill 2006).

A further aim of this research has been to investigate the difficulty experienced when legislation was initially sought to be established in the quarrying industry, and to show how the laws became more exacting throughout the period of study to the present day. Changes in the laws, within the quarrying industry, encompassed recognition of the need for improvement in working conditions in the 19th century, to environmental concerns of the late 19th century. These concerns included an awareness of the local population who were affected by dust, noise pollution and latterly subsidence.

Did these concerns about quarrying operations continue to escalate as the 20th century came and flourished? If so how did legislation evolve to address the people's apprehensions about present and future quarry sites? Research at Shropshire archives and local libraries discovered documentation concerning quarrying operations and the local communities in the 19th century. How did the 20th century approach the issues concerning the local residents living near quarrying operations? An interview with the Shropshire County Council's Mineral Planning Officer highlighted the attitude of governments and council policies on past, present and future quarry operations. Parish councils are another resource researched and interviews with local councillors provided their local community's perspective on the effects of quarrying, such as in the case of Bayston Hill village, which has Tarmac Ltd operating on their doorstep.

The research will examine the social and life-style issues within the quarrying industry from the quarry owners to the workforce through the access of personal wills. Although each site was individually owned and worked, the processes of extraction of the minerals were similar. When examining the historical documents from the Shropshire archives the importance of the records relating to the Grinshill sandstone quarry were realised to be particularly significant. Ledgers concerning Grinshill sandstone quarries from the 18th and 19th centuries document what the sandstone was being used for, the cost of the stone, how it was transported and the workforce who moved the sandstone. Shrewsbury offers an ideal example of the different purposes to which Grinshill sandstone was put, for example: bridges, and buildings. However saying this, there was and still are many different minerals of equal importance quarried in Shropshire, which made, and still make, the county unique in the Britain. The limestone extracted from Llanymynech Hill, a major quarry site on the English Welsh border, was used in agriculture and for building lime. When reviewing the aggregates quarries at Bayston and Brown Clee Hills,

Bayston Hill is still worked today for greywacke, which is a very limited resource, while Brown Clee Hill is now a worked-out quarry. However, although these minerals were both used for road making their locations reflect and give example of Shropshire's diverse geological landscape (Lapworth *et al,* 1910 and Toghill 2006). The limestone from two of the former limestone quarries, at Lilleshall and Wenlock Edge, were used for industrial purposes and originally in agriculture.

The heyday of Shropshire's quarrying industry was in the middle to late 19th century, the time of the industrial revolution when the limestone at Lilleshall was in demand for flux in the ironworking operation locally. The entrepreneurs saw their chances and took them; they constructed canals, it was Lord Granville Leveson-Gower in the beginning of the 19th century who obtained the services of the Gilbert brothers who had the legal and creative knowledge to construct Shropshire's Donnington Wood Canal and railways on the Lilleshall estate. Technological improvements in the quarrying and mining industries would be seen across Shropshire in the 19th and 20th centuries. Grinshill sandstone quarries had the first gas powered machinery and were first to use Bristar powder in the extraction of sandstone in Shropshire. Bristar powder is a very environmentally friendly way of extracting minerals. In my research no other quarry company in Shropshire used this Bristar powder method of extracting minerals; one wonders why, when it is proven to be environmentally friendly. This procedure does not seem to have been promoted by Shropshire County Council, but Grinshill Quarry Company was requested not to change to any other method of extraction. Maps of the area under research show how the quarrying industry started and expanded through the two centuries to the present day. In some cases the maps show the former quarry sites after the quarrying had ceased. Key themes in applications made today to Shropshire County Council, Mineral Planning Department is the need to show plans for the restoration of quarry sites after mineral extraction has been completed. Planned restoration at Bayston Hill greywacke quarry is an example of this procedure, which is outlined in this book. This shows how the quarrying industry has changed from the 19th century to the late 20th and early 21st century. One can see that redundant quarries of the 19th century were left to be reclaimed by nature. The danger of quarries just being left after extraction of minerals can be seen at former Grinshill quarry sites.

The book shows what can happen when groups of people get involved to save, conserve and renovate former quarry sites; a prime example of this it at Llanymynech. However, it also brings to light the difference when the two former quarries are situated in different countries namely

England and Wales. It needed the approach of an independent person who could see the national and local importance of the former quarry sites at Llanymynech. Mrs Harriet Devlin, M.B.E, would approach and point out the need to work together for the conservation of the now scheduled monument. Figure 1.1 shows the location of the quarries being researched in this book, along with some of the key sites mentioned.

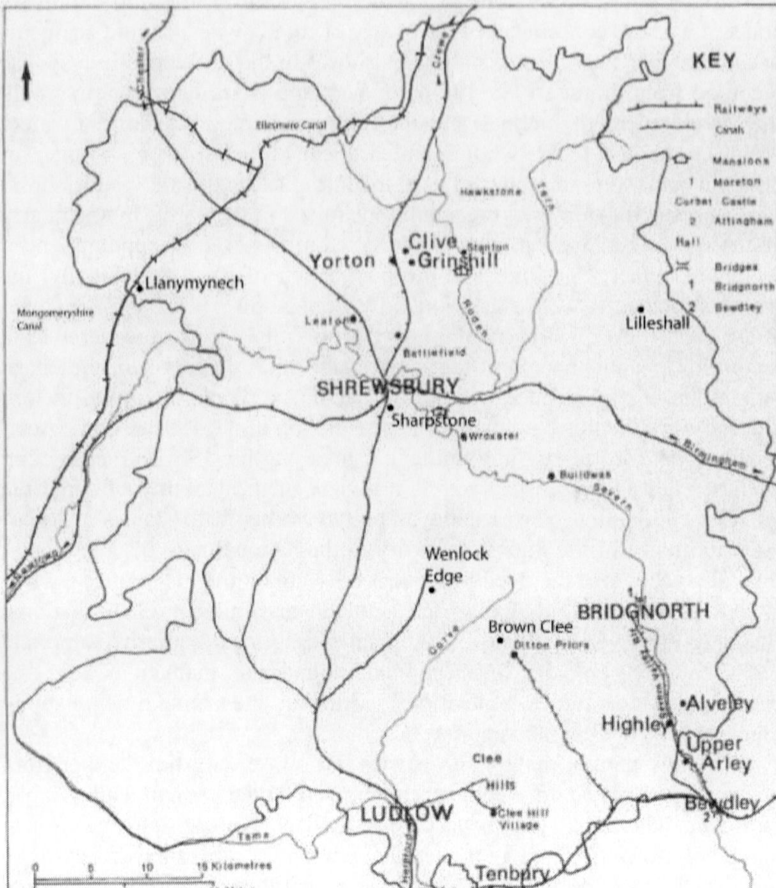

Figure 1.1: Map of Shropshire showing location of quarries studied and principal transportation infrastructure contemporary with the quarries. (Based on Scard 1989).

The research shows how quarrying companies introduced new blood to expand and develop their businesses. However, all these developments had long lasting effects on the local individual and family operated quarries, already affected by the Enclosure Act, 1818, which worked to benefit the entrepreneurs, rather than the individual operators. Further issues considered how safety was addressed and developed in the quarrying and mining industries during the 19th and 20th centuries in Shropshire. Of significant concern was the consequences for quarry or mine owners should accidents occur, whether fatal or not, and this is shown in the difference in response recorded from the quarry and mine owners and from their workers. With the development of inclines and the addition of new machinery, safety issues arose for those who operated them, whether men, women or children, who were unprotected until the introduction of safety legislation.

Concerns for safe working conditions raised in the 1830s by the quarry and mine workers were gradually addressed in 1842 by the appointment of a Commissioner for Mines and the publication of its report. Finally, the workers' concerns were championed by Lord Ashley with an amendment to the Factory Act, 1891, enabling open air premises to be considered as 'a factory'. Up until that point the quarry and mine owners had argued, at length, that an open space was not the same as a regular enclosed factory site and were therefore not under the jurisdiction of the 1891 Factory Act.

Reasons for quarry and mine closures in the 19th and early 20th centuries are addressed, as well as a review of the viability of some the quarries continuing to work today. The government of today is aware of the importance of the mineral industry to the national and local economy; they also recognise the need to maintain a viable minerals industry, while at the same time protecting the natural environment. The National Planning Policy Framework, 2012, enables each local authority to protect both the safety of the local population and the maintenance of the environment for future generations, whilst at the same time enabling viable quarry working to continue.

It is only natural that some quarries will close over time, either from becoming exhausted, or when extraction is no longer economically viable. However, when quarry sites become redundant they are left to return to nature, or are landscaped by the quarry company. At this time there will be the development of local, specific fauna and flora, some of which are uniquely associated with the exposed underlying rock. As a result, certain areas within the former quarry sites may be designated as Sites of Special Scientific Interests (SSSI), which will need to be taken into account in any future plans for the restoration of the area. The SSSI designation enables the legal protection of these unique and limited areas. In towns and cities former factories and their machinery may be considered of national or local importance and are listed in order to protect and acknowledge their contribution to the industrial revolution of this country. It is important to recognise the significant role of Shropshire's quarrying industry in the construction of bridges, buildings, factories and railways across the country. Therefore, more should be done to conserve, not only the fauna and flora SSSI, but also some the former quarries and machinery, a prime example of this is the moth-balling, in 2007, of Lea Quarry site at Wenlock Edge, and its subsequent complete clearance. It is now a timber yard; nothing is left of the old buildings. Llanymynech is an example of another similar quarry site which was scheduled on the 14th July 2006 under the Ancient Monuments and Archaeology Areas Act 1979. Llanymynech is of archaeological and historical interest and so of equal importance as any sites and artefacts from Neolithic or Roman period. These former quarry sites could then become heritage places of local and national interest, by relating and expanding on the knowledge of the archaeological, historical and human past.

At the end of the 19th and early 20th centuries, the quarrying industries started to go into decline. Research indicates that causes of the deterioration in the minerals industry were related more to a labour shortage than to that of economy. With the coming of World War 1 in 1914 the quarry labour forces were needed as soldiers, and the country's economy was directed to the support of the war effort. Even where, at Llanymynech in 1899, a new highly efficient kiln had been installed, the whole quarry was forced into closure by 1914.

1.4 Field Visits

All the quarry sites researched in my book were visited to achieve and gain knowledge from the still working quarries, which would hopefully provide a better understanding and perspective on the redundant quarries. By looking at redundant quarries reclaimed by time and nature, it was possible to envisage what the future would be for the quarries that were still active. Lilleshall and Wenlock Edge Quarries, both redundant, were the first sites of the six researched. Lilleshall Hill Quarries are located northeast of Lilleshall Village. Wenlock Edge Quarries are located on the B4371 the Much Wenlock to Church Stretton Road. Grinshill Quarry was the next quarry to be researched. It is a working quarry which lies approximately 7 miles northeast from Shrewsbury on the A49. This quarry contains creamy white and light reddish colour sandstone. Llanymynech Limestone Quarries, now redundant, was the third site researched and lies approximately 5 miles southwest of Oswestry. Bayston Hill, greywacke, and Brown Clee Hill, for dolerite, quarries were the next sites researched. Bayston Hill Quarry is a working site, backing onto the outskirts of Bayston Hill village, located 3 miles south of Shrewsbury. Brown Clee Hill quarries are a redundant site located near to Clee Village on the B4364 Bridgnorth to Ludlow Road.

The former Lilleshall quarry sites were the first to be visited on the 22nd August 2009. On this visit photographs and notes were taken of the canal and former wharf where minerals were loaded and unloaded. In addition, at Lilleshall, access roads and bridges were noted and the landscape of the former quarry's sites was visited to see what it looked like today and how nature had reclaimed the mines and quarries site now.

The site of Wenlock Edge's former quarries and surrounding areas were visited on 14th April 2010. Photographs and notes were taken of the former quarries and surrounding landscape, which included the former railway. Former quarry machinery such as the primary crusher, hoppers and conveyor sheds of Lea and Coates Quarries were examined. Buildings at Coates Quarries had been dismantled, all excepting for the primary crusher, to create a storage area for fencing and other items. Talking now to the former workers of the closed Coates Quarry was interesting because they said now that they import limestone from France to use when required. In addition, I walked up and onto Wenlock Edge where one could get an overview of all the former quarry sites. Looking from this vantage point one could appreciate the amount of limestone that had been quarried within the vast area, as well as see former individual quarries once worked by one person or a family. The site of the former Wenlock railway could also be seen. The disused redundant quarries are now being reclaimed by the natural elements, animals and plants. While standing on Wenlock Edge fencing can be seen, but it would appear to be inadequate to prevent the general public and walkers from falling over the edge. The dangerous depth of the former quarries from such a height could cause fatal injuries to people.

On the 8th November 2011 I met Trevor Kirkham who gave me a tour of the Grinshill Quarries, which lasted all morning and I was able to take photographs and notes. Trevor firstly took me around the working quarry site and then the redundant quarries sites. He informed me of the archaeological and historical background of today's and former quarries. We looked at the ashlar blocks that had been extracted by the Bristar powder method; Trevor went on to explain the benefits of this powder. Then we also walked around Corbet Wood where old, individual redundant quarries could be seen and the former site of quarrymen's cottages had been. In the interview Trevor talked about the future of Grinshill Quarry. I asked about the present owners, Mr John O'Hare and his wife. I asked Trever if he thought it would be possible to talk to the new owners, but he did not hold out much hope of obtaining an interview with them. I did telephone asking for a formal interview which was declined without a reason.

After undertaking my research on Bayston Hill Quarry, I decided to do a site visit on the 29th March, 2011. A Public Right of Way runs straight through Bayston Hill Site from the entrance of the processing area and around the outside of the redundant and working greywacke quarries. The Public Right of Way was altered to go around the outside of the quarries. This footpath now provides an excellent view point from which to take photographs and make notes of the quarrying operations. I was able to see the new quarry from this vantage point and in particular how the newly quarried greywacke was transported from the face at the base of the quarry by dumper-truck via constructed roads that go around the quarry sides and so taken to the processing area up above.

From the Public Right of Way, I was able to see where trees and shrubs had been planted on top of a mound to help screen the quarrying operations. I took up different points along the Public Right of Way to give me different views of the quarry. Fencing had been constructed around the working quarry at a safe distance, which I observed from different points. I also noted the safety notifications and warning of blasting times. Walking back through the processing area, byway of the Public Right of Way, I was aware that the dust from the processing sheds was quite visible and had thickly covered my car. By this observation I wondered about the working conditions within and around the processing sheds. I could see that the workingmen had safety clothing on, but how much of the dust did they keep out? In addition, I could understand these safety points from the viewpoint of the Bayston Village Liaison Committee. However, the Committee seemed to have good working partnership with Tarmac Ltd who were constantly cleaning the in and out access road to the site. Llanymynech was visited on the 4th April 2013 where I met Mrs Joan Zorn. I was able to take photographs and notes throughout my tour. Mrs Joan Zorn showed me around the former horse stables and the tracks to the two canals and wharfs. From there we went into the Visitors' Centre which contains information boards on the working history and ongoing restoration funded by the Heritage Lottery Fund and undertaken by the Friends of Llanymynech. Also within this building was a model of the Llanymynech Limeworks. She showed me the former incline tracks within the woods and throughout the lime works, as well as the English and Welsh Drum Brake Houses and she informed me of the work being proposed for the English Drum House. From the top of Llanymynech Hill, I was able to see both the English and Welsh quarry inclines and the distance between them. Joan also showed and told me about the Tally House and its working, where the quarried stone was weighed and recorded.

At Llanymynech, it was quite obvious to see the differences between the ownership of the English and Welsh quarries. There were two separate tunnels under the road, one belonging to the English and the other to the Welsh owner, with their own inclines and tramways from the separate quarries. Joan also told me of the lime kilns and the introduction of The Hoffmann Kiln and how it worked and was loaded with limestone blocks, as well as about the conditions and daily work of the quarrymen who loaded the lime. I was given a tour around the Hoffmann Kiln inside and out. I was impressed by the amount of informative boards placed at many strategic points all around Llanymynech Limestone works. Later I was informed by Mrs Joan Zoan of the historic background of both the English and Welsh families that owned the quarries and the differences between them.

This introduction provides some of the key themes and approaches addressed in my book on the importance of the quarrying industry in Shropshire. The minerals industry of the county has played and continues to play a part in the economics of Shropshire. Today the quarrying industry is not as large as it was in the 19th and 20th centuries; however, Bayston Hill Quarries are still extracting greywacke mineral. At the same time redundant quarry sites can now be seen in another light, where the ecology of returning fauna and flora can be studied. The redundant sites provide an insight into the development of industrial machinery and the working and living conditions of Shropshire's past. At the same time these sites can now be seen in another light, where the ecology of returning fauna and flora can be studied.

CHAPTER 2

LITERATURE REVIEW

2.1 Introduction

This chapter outlines the principal research themes of the book thus outlining the current state of knowledge. The first section will cover the diverse geology of Shropshire and its relationship to the quarrying industry enabling different products to be extracted such as dimension stone, aggregates hard-core and setts which were used locally and across the country. I then move on to give an idea of the quarrying industry as a whole; starting with a brief review of early technology, following on to larger scale quarrying as the industrialised era progressed, because the past, present and future contribute to the entirety of our understanding.

The book will examine and show the limitations in previous research, filling in the gaps left in the history of mineral quarrying in Shropshire in the 19th and 20th centuries. The gaps in the research are significant and therefore it is not possible to gain an all-round understanding of the quarrying industry and its social impacts on the workforce within Shropshire. In particular, I will consider how and why demand on the quarrying industry altered in the middle of the 19th century. Lastly, there is an examination of the legislation relating to the quarrying industry and its impacts upon the continuation of quarrying within the county.

2.1.1 Specific case Study

This case study was chosen because there has not previously been a detailed research showing the importance of the bulk mineral quarrying of Shropshire between 19th and 20th centuries. It will provide examples taken from Shropshire's bulk minerals quarries, together with its geological diversity. Shropshire's geology is very diverse and most rock periods and types can be found there. I acknowledge that there are sand and gravel quarries; however, it would be impossible to research both aspects of the Shropshire quarrying industry within the proscribed word limit and time. The approach has been to provide examples of each aspect of the hard-stone quarrying industry within Shropshire from both working and redundant sites. Shropshire is the largest landlocked county in England. The county provides the greatest opportunities for mineral extraction, while at the same time presenting problems for its transport. These sites were selected because of their resources, their growth and economic importance to Shropshire. The study will show the contribution of these minerals to the industrial revolution as well as Shropshire's infrastructure, such as roads, bridges and buildings.

2.2 Geology

Shropshire is considered to be a place of outstanding natural beauty; its hills cover more than a quarter of the county, especially The Shropshire Hills, Wenlock Edge, Long Mynd and the Stiperstones, many celebrated in Houseman's poetry.

Shropshire is mainly rural and sparsely populated when compared to other counties (Lapworth et al, 1910 and Toghill 2006, pp 123-52). Peter Toghill's book provides an up to date assessment of the geology of Shropshire; however, Lapworth et al, 1910, still contributes useful information.

The geology of Shropshire is so diverse that most geological eras and types of rocks found across Britain are significantly represented in this county (Toghill, 2006). Knowledge of geology identifies the location and quality of mineral resources so that quarrying can take place; thus the two disciplines are linked together. It is accepted that in Shropshire there are ten geological periods, whereas, throughout the country there are only twelve in total, so this makes Shropshire very unusual and distinctive (Toghill, 2006, p27 and Stanier, 2000). As Toghill remarks, *"This is remarkable when one considers that in an area like Snowdonia, only three geological periods are present."* The map above (Figure 2.1) displays the different geological periods giving the intrusive and extrusive rocks: andesite, basalt, and dolerite, all of which are of interest to the Shropshire quarrying industry.

There is evidence that the very diversity of rock types and periods influenced the location and development of many of the early industries in this region. It contributed to the construction industries and infrastructure of Shropshire as well as to the country (Trinder, 2000). In the 19th and 20th centuries, quarrying in Shropshire became a major industry; however considerable quarrying still takes place today. In Shropshire the varieties of stone available to builders, road makers (setts) and agriculture, gave Alec Clifton-Taylor pause for thought to wonder whether any other country, as limited in area as England, might be able to offer so wide a choice in the matter of construction materials (Clifton-Taylor, et al, 1994).

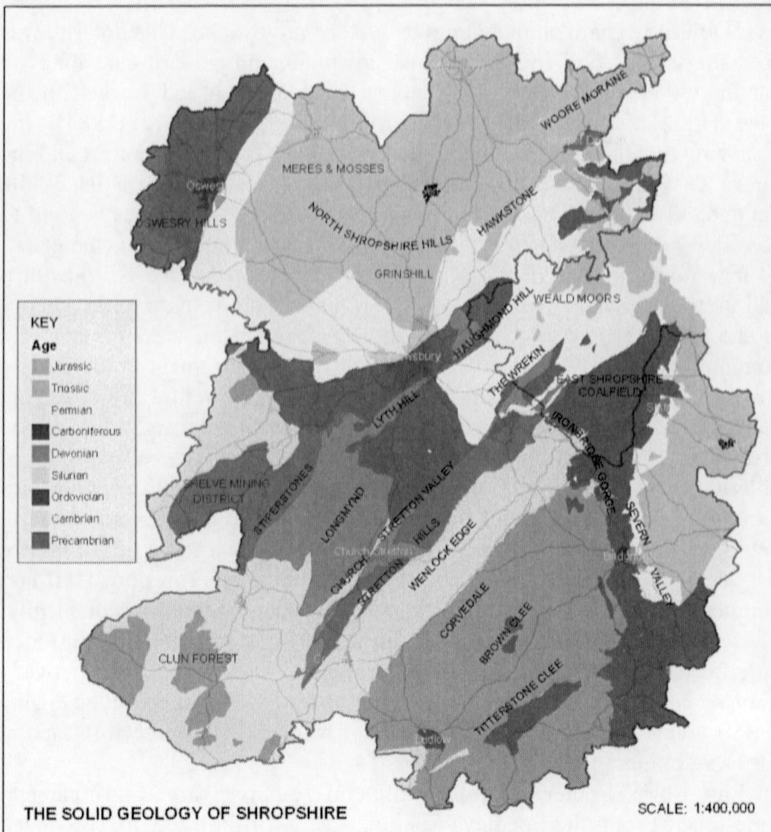

Figure 2.1: Source: Toghill, 2006. Geology of Shropshire. (Shrewsbury, Swan Hill Press).

Shropshire was the birth place of the Industrial Revolution, being a rural region made the minerals more easily accessible whether quarried or mined; these sites are now being reclaimed firstly by nature and in some cases by Historical Societies. In geological terms the oldest rocks in Shropshire are located on Bayston Hill and Haughmond Hill near Shrewsbury; these rocks are of the Precambrian period. At Bayston Hill these sedimentary rocks, industrially known as greywacke, are quarried and processed for making roads (Toghill, 2006). At Brown Clee Hill, the Igneous rock, dolerite, can be found on Brown Clee and Titterstone Clee Hills in South Shropshire. The summit of Brown Clee Hill was capped with Dolerite, a hard stone but not as hard as greywacke. This dolerite was used for setts to cobble streets, canal towpaths and as hard-core for road making. A local historian, Alf Jenkins, who lives on and worked in the Clee Hills, has chronicled the story of his community (1983); the quarrying and the men who worked at the quarry face, their dialect and the trucks as well as about his life and growing up on the Clee Hills. While undoubtedly a useful social and local history, this study moves beyond to consider the wider nature of quarrying in Shropshire and how it changed.

The northerly neighbour of the Clee Hills, Brown Clee Hill has also had its local historian, Di Bryan (2006). She sought to narrate the history of the hill, the quarrying and the railway through the view of the local community in the small town of Ditton Priors, a settlement at the foot of Brown Clee. Her book focuses more on the history before the quarrying and associated industries, but saying this it gives a locals' view of the history of Brown Clee Hill. The quarry on Brown Clee also produced dolerite, as on Clee Hill and Titterstone Clee, as well as Old Red Sandstone which was found underneath the dolerite on Brown Clee Hill, being considered as one of the finest resources of road stone in Shropshire (Bryan, 2006 and Toghill, 2006). The quarry at Bayston Hill and neighbouring Haughmond Hill has never been studied academically. These quarries produce greywacke, an earlier rock type than the intruded dolerite. Although similar in appearance, the difference between greywacke stone and dolerite are their different Polished Stone Value (PSV) levels (British Geological Survey, Minerals Resource Information for Development Plans, 1998).

Shropshire's other hard-stone mineral resources are sandstone and limestone. From the medieval period, the Grinshill Quarry produced dimension sandstone which was very much in demand for building castles, houses, schools although output is now limited providing stone for building restoration. Llanymynech, Lillishall and Wenlock Edge Quarries all produced limestone for various purposes at differing periods: agriculture, fluxing for furnaces, aggregates and building stone (Toghill, 2006). None of these quarries has been seriously studied except for Grinshill (Thompson 2004a & b; 2005, 2006; Thompson and Lillistone 2008). These privately produced studies do offer an account of the quarry industry but they are not easily accessible.

2.3 History of Quarrying in Shropshire

2.3.1 Literature, technologies and start of large scale works

A study of the pre-industrial mining and quarrying industry in all its forms has recently been undertaken on behalf of Historic England; this includes other major quarried minerals such as Bath and Portland Stone (www.historicengland.org.uk, accessed 17/10/2016; Topping 2011a & b; Newman 2016). A brief analysis of how the foundations of the early quarrying industry started before the industrial development is important to undertake in order to show that, without the first peoples' innovations new technologies would not have come into realisation. The main focus of this literature review will be on the changes in technology adopted in the Shropshire quarrying industry from the 18th century, rather than focusing on the earlier periods. In this way, the impact of new technologies can be traced, alongside the development of the industry and supporting infrastructure.

It is clear that people have always understood, and exploited the properties of minerals; this is demonstrated by the remarkable quarrying and transportation of specific stone for Stonehenge from the Presseli Mountains in South Wales (Parker-Pearson, *et al* 2015). Equally it is also clear that early people exploited the materials they found on the surface, as with the Stonehenge and Avebury Sarsens (Chippindale, 2012, pp.38-40). In Shropshire, similar activity can be seen on the Brown and Titterstone Clee Hills where surface rocks would have been found and used to construct the ramparts of hillforts and other structures, such as houses (Marriott, et al, 2008). Working the stone was limited as tools were basic, such as deer antler, but it is clear that there was an understanding of the natural rules of bedding and the benefits of weathering (British Geological Society, technical report, Shropshire: resources and Constraints, WF/98/6, 1998).

In the Roman and Medieval periods, understanding of quarrying technology and the identification of suitable rock types improved. Specifically, there was an application of new and stronger metal-based tools, allied with a deeper understanding of what made certain rock types suitable for building. Throughout this whole period, however, stone masons continued to rely on basic tools and technologies, not having access to gunpowder for instance. These factors undoubtedly acted as a constraint on quarrying ability, although feats such as Hadrian's Wall and the great medieval castles and cathedrals make it quite clear that much could be achieved with this existing technology (Parsons 1990 and Stanier 2000).

A major development in the early modern period was the impact of the Reformation which made available considerable assets in land to entrepreneurs, as can be traced in the development of Coalbrookdale (Alfrey and Clark 1993). Entrepreneurs like the Darbys 'seized the opportunity' and purchased vast parcels of monastic lands, which lead eventually to the development of technological advances. Firstly, at the beginning of the 18th century, this occurred with Abraham Darby's discovery of coke which he used to increase the temperature in his blast furnaces for iron smelting, boosting production and the need for limestone as a flux; this continued with Abraham Darby IV. In the 19th century, the Darbys, and other iron-masters, were prominent players in the iron industry of Shropshire which was a major element in the industrial development not only of Shropshire, but of the whole country. The iron industry relied on basic raw materials found in Shropshire, such as limestone and coal for making coke (Trinder, 2000; Cossons, 1993). However, in the 19th century there were other minerals, such as coal, stone and clay, besides limestone that were quarried, which also assisted in the growth of the manufacturing industries. The quarrying industry was as old as time, however, as industrial growth took hold in the 19th century, changes were afoot that increased the scale of production and advances in technology. With industrialisation, there was an increased demand for bricks, limestone for mortar, cement and in time concrete, which in turn injected growth into the building industries. The same impetus led to the industrialisation of agriculture and the need for artificial fertilizers and soil improvers, such as limestone and basic slag. This increased the need for more craftsmen across all the fields of industrialisation. Contemporaneously, new infrastructure developed in transportation, such as: turnpike roads, inclines taking the stone down to canals, and eventually railways which would replace the canals, as in turn lorries replaced the railways, thus advancing the quarrying industry into the 18th, 19th and 20th centuries. The entrepreneurs who grasped the opportunity to develop the former monastic lands were also the first to build a canal in Shropshire; this was that built at Lilleshall, the Donnington Wood Canal (Yates, 2008 and I.B.G.M. L, Ove Arup *et al,* 1987). They would eventually build stone dressing; saw processing machineries; conveyors, crushers and grading sheds in the quarries (William, 1997). However, as with everything, in time, what had once been transported by rail, then lorry, now in the 21st century, will return to rail transport which has re-opened at Bayston Hill Quarries to again convey the quarried stone across the county and the country (Trinder, 2000).

These latter developments demonstrate the effect of current government focus on environmental issues, such as transport, congestion, vibration, noise and the natural environment. Grinshill Quarry is a prime example of the above issues because of its location. The site is surrounded by a wood and has limited road access. In the 19th century, it supplied most of Shropshire with its building stone until the early 20th century when operations ceased for a time, then started operating again at a much smaller capacity. The quarries never again reached the pinnacle of production experienced in the middle and late 19th century. Today the quarry just supplies stone mostly for restoration. Even though production at Grinshill Stone Quarry is greatly reduced, this quarry is the only one in Shropshire to use the most advanced explosive, Bristar Powder, in the extraction of stone blocks, as it has the most minimal environmental effect (Clifton-Taylor, *et al,* 1994).

Clifton-Taylor's book only mentions one quarry in Shropshire, and even then only in a small paragraph. This has been the constant problem when researching literature on the quarrying of Shropshire in the 19th and 20th centuries. Although the industrial archaeology and heritage or Shropshire has been widely studied (Downes, 1980; More, 1997 and Trinder, 2000), quarrying has been a strangely neglected industry discussed in only brief studies. An exception to this is the Nuffield Survey of the Bentall Edge lime quarries and the Lincoln Hill galleries, which were in the past a tourist attraction a bit like the Bath stone quarries near Bath (Alfrey and Clark 1993) and the article published by Scard (1989).

In Sir Neil Cossons', 1993 book for example, still a highly regarded work for the discipline; there are only three pages on quarrying stone. His book refers to the momentum of the industrial revolution during the middle of the 18th century and all of the 19th century. It talks about the rapid evolution of industry and innovation that took place, but the industrial growth of Britain needed natural minerals such as limestone, sandstone and coal for furnaces. That produced the iron for railways which were the lifeline of the industrial revolution; he talks about the canals, bridges, roads, and the factory that was the first iron-framed building, constructed in Shrewsbury. Although, Cossons' book only deals directly on quarrying stone in only three pages, it can be said that without quarrying for minerals the growth that evolved with the industrial revolution might not had happened because the quarrying for limestone helped in the making of iron in the furnaces of Shropshire. This in turn contributed to making the iron railway tracks and other constructions that needed iron frames. Trinder, (2000) like Cossons, writes on the subject of industrial archaeology; however he focuses on Shropshire, which provides a better understanding of momentum that took place in the county. His book talks of the growth of Shropshire's industry from the horse-trains, tramways, canals and railways which transported merchandise around the county. Tinder's study briefly refers the quarrying and the minerals that can be found in the Shropshire, the innovation such as Donnington Wood Canal the first canal to be built in the county by the Lilleshall Company who quarried limestone. In addition, one can see how the industrial revolution took hold in Shropshire from the ironmasters of Coalbrookdale, the River Severn and how the minerals were used across Shropshire in different industries, for example aggregates and concrete; to the decline in manpower after the First World War. This shows that in just four cases how limited any reference to quarrying is, in any form, so these books cannot provide an in-depth analysis of this very important industry. More specifically, Downes, 1980, Industrial Archaeology, is the example of a book, which refers to a number of quarries across the country, but only has a single paragraph about Grinshill quarry; this was a common occurrence when researching for my book and was the principal reason for undertaking the research reported on here.

The Quarry Manager's Journal, 1968 and 1972; I found to be a good source of information on the quarrying industry, not just for the quarries I researched in Shropshire, but for the whole country. They were journals written by the quarrying industry itself, which gave a unique understanding of the process of the quarrying from their point of view. They wrote about the technical developments and men working within the quarrying industry. However, the journals only focused on their own industry and the developing industries that needed their materials.

For an analysis of quarrying in Shropshire, virtually the only article is that written by Ann Scard (1989) who writes on 'The Development and Changing Organisation of Shropshire's Quarrying Industry: 1750-1900'. She comments on the diverse geology of Shropshire and the different expanding markets, which are developing for the building and engineering stones of Shropshire. Scard talks about the importance of the small quarries in the 1750s and their indispensable value to the scattered communities of Shropshire. How the local building projects involving engineers of national importance. Scard goes on to write about the developing quarrying industry from the small quarries to the large commercial quarries and the technical innovations that took places over the hundred-fifty years of the industry. She notes the development of transport infrastructure from the early waggons and horses, tramways, canals, to railways and lorries. She also comments on the working traditions of the quarrying industry from the early manual methods, which included men, women and children. The methods of working continued within the quarrying industry despite the growing demand for quarried stone whether freestone or road stone. Scard writes that the changing times and the developing architectural and engineering styles within the construction industry which began in the 19th century and would lead to a decline of the traditional quarrying methods. However right this may be, everything has to develop and advance so that they continue to progress to a more efficient industry, which is the quarrying industry today in the 21st century.

Scard writes about the period of 1750 to 1900 around the start of the industrial revolution in Britain, but the period leading up 1750 and after 1900 is of equal importance. The quarrying industries were developing before this time and have certainly advanced sine the 1900, which can be seen today in the 21st century. The usage of stone in the period under discuss would not have intensified if the foundations had not been laid before this time. Scard acknowledges how stone was as important in construction of buildings and bridges, but the potential of the stone quarrying industrial would not be reached until the industrial period of this country. She talks about the other materials that were being used in construction such as, bricks and iron, however there were other materials invented by the industry like concrete and imitation stone. Scard believes that these materials were no overall threat the stone quarrying industry. However, she says these new trends did replace the intensive labour for natural stone in the 20th century. She talks about the modern quarries in the 20th century; today the stone is blasted from the quarry face not like it used to be done by manual labour. The article does say that there are a few small quarries that are still operating for the sole purpose of renovation in Shropshire. Between the years under discussion she says there was a combination of different industries and agricultural interests. The cities and towns that benefited from the growing stone market in this country, which came about firstly with the coming of the canals and then the railway, made Shropshire quarrying mineral industries available to the whole country and beyond. Scard again notes how the building and engineering expanded considerably between 1750 and 1900.

Scard seems to focus on Grinshill, The Clee Hill, Highley and Alveley Quarries, these were the main quarries in the 19th century. While I acknowledge their importance there are many more quarries of equal importance to the quarrying industry of this county and of the nation in Shropshire. Other quarries I consider to be of equal importance are for example Wenlock Edge, Lilleshall, Bayston Hill and Llanymynech Quarries since through the different minerals they produced; they contributed enormously to the industrial revolution and growth of Shropshire and the Britain.

A different perspective on the time and industrial practice is found in Gallop's study of children in Victorian times who worked down mines (2010). He writes about the hardships of the Victorian children who not only worked down mines but in factories in the 1830s. Children as young as five years old were taken down mines; they worked as trapdoor keepers they worked between 10 to 14 hours per day. This author also writes about an accident in 1838 which killed 26 children who at the time were working down the coal mine; this accident would have far reaching effects on health and safety, child labour across all working areas from factories, mines and mills. The main instigator of the change in child labour legislation was a Lord Ashley who chaired the Royal Commission. Lord Ashley was responsible for banning girls and boys under the age of 10 years old working in mines; this came out in the Children's Employment Commission report of 1842. However, health and safety of the men, women and children that worked in the quarries seems not to have been taken into consideration, although the cause Lord Ashley campaigned for would in time affect all industries in England.

Marriott, *et al,* (2008), is a fairly new assessment on the importance of conservation of a site of not just local, but national importance at Clee Hills in Shropshire. This reports aims to bring to the attention of the appropriate people who are in a position to put into place the conservation plans that are outlined in the documents. The report writes about the historical and archaeological remains on Clee Hills and the need to preserve them for the future such as the hillfort and the quarry remains. The report also highlights the importance of the different ages throughout time on the Clee Hills from the prehistoric to the industrial landscape today. The Clee Hills today has tremendous educational significance from every period from the Neolithic, Romans and Medieval. The authors suggest in the report that the archaeological potential is still to be appreciated. The industrial activities on Clee Hill makes it very important in industrial terms, with dolerite quarries, coal mine "bell-pits". It also writes on the ecological, geological and aesthetic aspects of the Clee Hills. With all the activities that have taken place on Clee Hills the report writes of the social and community importance, which goes with all periods of time on the hills; without people the community and the artefacts would not be found today to study. All of the archaeological sites on Clee Hills, whether buildings, former quarries, cairns and artefacts need to be legally protected not only from the people, but also the natural elements. It talks about the Clee Hills, however today there is only Titterstone Stone quarries still working, but Brown Clee in its day was of equal importance. It summarises the immense importance of the Clee Hills and its past and what it can give to people of the present and future.

I acknowledge that some books, journals and reports were researched and written by individuals, which focused on particular sites such as Thompson who wrote five reports on Grinshill Stone Quarries covering its history from the middle of the 16th century to 20th century. These reports are unique on the historical and archaeological background of Grinshill Stone Quarries, from their connections with buildings in Shrewsbury, the English and Welsh Bridges to a whole range of buildings all over Britain. The report covers the heyday of the quarries from Victorian times to today when the quarry site is very limited in resources.

Stanier writes about the stone quarrying landscape of Britain from the time when he was growing up when he lived in a Cornish town, surrounded by granite and slate (2000). He comments that even at a young age that there was a sense of a declining trade, because there were fewer men working on sites and there was an air of negligence; empty timber sheds. He acknowledges that quarrying is as old as time itself. From surface stone to holes in the ground, stone was quarried by early man with antlers and stone tools. He refers to the construction of stone tombs, circles, standing stones in Neolithic, Bronze Age and the coming of the Romans.

Stanier comments about the Roman quarries, around Bristol, the tools they used to quarry stone and their structures that can be seen across Roman Britain, particularly in Bath, Somerset. He also goes on to talk about all the different periods of Britain's stone industry and how the minerals were quarried and used right up to the modern times. One of his observations is that most of the quarries seen in the landscape of Britain today are of the industrial period.

He writes that almost every type rock that can be found and quarried in Britain has been used in buildings, bridges, docks and road stone etc. He discusses the famous Jurassic belt from the Dorset coast through Somerset, and another four counties to Yorkshire. To the Triassic sandstones which can be found in Shropshire and other parts of the Midlands; the importance of local geology particularly in Shropshire. He also confirms that rivers were major sources of the transport in the Middle Ages.

Stanier notes that quarrying is not featured very much in the archaeological records; he says that there has been more written on stone masonry. He writes on the different stone types such as millstones of the Peak District how it was quarried and worked; today some of these millstones can be seen still in the landscape with tool marks on them. He also refers to the importance of Bath and Purbeck stones and their quarries, which in some cases are still working today.

In his book can be seen references to old former quarries sites of the late 19th century and quarrymen and stonemason of Purbeck quarry. This book also writes of the different locations of quarry sites some on cliffs on the Isle of Purbeck. Blasting had taken place in quarries, so machines such as cranes would move the quarried stones in horse drawn carts and in time waggons. Stanier also refers to the concrete structures left today on the Clee Hills, remnants and archaeological records of the industrial past of the hills. He writes too about different types of transport used in the quarrying industry from the man, the horse, rivers, inclines, bridges, turnpike roads and roads.

Stanier summarises the importance of lime kilns, but there is little mention of quarries that fed the kilns form the earliest period to the present time. He records the minerals that have been extracted from the landscape across the country over time and those that are still being quarried today; he enumerates the minerals that have been used in buildings, bridges, monuments and in the construction of the roads. Stanier comments on the effects of quarrying on the landscape about which one can raise the question as to whether these sites should be considered as scars or just holes in the ground. Even after being reclaimed by nature, the significance of these quarries is in their contribution to the archaeological record of Shropshire and elsewhere.

Today these old quarries are considered to be of local and national importance to geologists and biologists, among other scientific disciplines, who wish to study the plants that return on the once quarried land; birds and all creatures that return. Additionally, these are good places to walk in the natural landscape and view with care, so gaining an understanding of the industrial, archaeological and working people of the past who made this country's infrastructure. This can be seen in The National Stone Centre, Derbyshire, which is an ideal place to start if one wants to understand the stone industry and the people who lived and worked in quarrying.

These authors have made significant contributions to the study of the quarrying industry of the country. However, when studying the country as a whole, there are naturally limitations and gaps in research into documents and knowledge specific to each county. Some authors write about the Shropshire quarrying industry, but only briefly so that it is not so easy to gain a wide breadth of understanding of the county's mineral industry. My research is the first comprehensive study of the Shropshire quarrying industry, the pre-eminent industry of the county. Shropshire's quarrying industries' contribution to the industrial revolution of this county and the country was tremendous. In Shrewsbury, the county town, examples of the beauty of the stone can be seen in the Welsh and English bridges, the railway station and the grammar school. The uniqueness of the bulk minerals brought entrepreneurs to the county to quarry these minerals; they were drawn by the distinctiveness of the Shropshire's geological makeup. As has already been mentioned, ten geological periods are found in the county as compared with other parts of the country. Toghill, (2006), and Scard, (1989), mention this but no other authors have considered this geological importance. I would have thought this is a major point in any discussion on the quality of the minerals of the county.

Shropshire is considered to be the birth place of the industrial revolution; many authors have highlighted this with regards to the ironmasters of Ironbridge. The quarrying industry of the county has only briefly been discussed, yet without such a mineral as limestone, iron making would not have happened. It is equally important to understand, the process of extraction of bulk minerals, the contributions of the workforce, technological developments, the minerals' usage in construction, ironworking, as well as in agriculture, to gain a complete picture of the quarrying industry of Shropshire.

Quarrying in the 19th and 20th centuries was a major industry in Shropshire. , Even before this period quarrying for minerals was a way of life for individual people and families. The importance of Shropshire's quarrying operations go back a long way from monasteries, farmers, individuals, families, to entrepreneurs who developed the quarrying industry. This early industry was supported by an infrastructure of pack-horse tracks, canals, railways and road systems across the county and Britain. Shropshire is one of the most rural counties in England; apart from Shrewsbury having only a few small towns. Initially the River Severn, which runs through the county, was the main way of transporting quarried minerals. One must remember that agriculture was the major industry before quarrying came to its peak in the 19th century. Up until the Industrial Revolution, limestone in the county was primarily used to provide mortar for building and as an improver of agricultural land.

Other research has not considered in fuller detail what the redundant quarries still offer today as places for the recolonization by fauna and flora, while at the same time they do acknowledge the industrial importance of the mineral industry. Even working quarries today have to take into consideration the effects of quarrying on the environment and wildlife, while they are still employing local people and contribute to the local economy. Although in the 21st century the quarrying industry has altered, it still has a social impact.

2.4 Key Aims of This Research

This section outlines the key aims of the research:

- To explore the significance of the Shropshire quarrying industry.
- To understand the social impact of the quarrying industry on the workforce of the region.
- To explore the historical background of the individual quarry sites, so as to provide an understanding of their industrial growth.
- To investigate 19th century legislation and its applicability to the quarrying industry up until the present day.
- To record the technological developments on quarry sites i.e. tramways, inclines.
- To follow the innovations made in transport of quarried minerals by pack-horse, canal, railway and lorry.
- How changes to quarrying were established to protected archaeological finds, the environment, ecology, conservation of the flora and fauna.
- The present day approach to the restoration and legal requirements of after-care of redundant quarry sites.

2.5 Research Methodology

When starting my research into 19th and 20th century quarrying in Shropshire, I firstly set out a plan of action: located the different quarries that I intended to investigate; established whether they were still working, or redundant; what minerals had been, or are being, extracted; whether these are limestone, greywacke, dolerite or sandstone. In addition, I was able to enhance research drawn from archives with interviews with people associated with the industry and the land. It was possible to read papers held in the Collection of the Ironbridge Gorge Museum Trust showing that, following the Dissolution of the Monasteries, James Leveson, a member of a wealthy wool trading family of Wolverhampton, was able to purchase the Lilleshall Abbey lands on the 24th December 1539. Maps and wills of c.1780's, held in Shropshire Archives were accessed to establish and identify the location of individual family operated quarries. Company records and ledgers held in various locations including at the Staffordshire Record Office, Shropshire Archives and the National Library of Wales, record the effect of 18th and 19th century entrepreneurs on the quarrying industry in Shropshire and the technological developments they brought about. During the 19th and 20th centuries, developments in technology brought new hazardous working conditions, only some of which were protected by legislation. During the 19th century, research revealed how conditions of work were applied to the continued lease of housing to quarry workers employed at the Lilleshall and Brown Clee Hills quarries.

2.5.1 Quarry Research contacts

At first I constructed a plan of research to find out whom if anybody was or had done any studying of the quarries sites in Shropshire. Visits, telephone calls and emails were sent to local and county libraries in Shropshire; letters were sent and placed in the local edition of Bayston Hill's *The Villager*. Through the village magazine one contact was made, but he was not able to assist. There was a local archaeologist, Ernie Jenks, but because of his age and poor health he was not willing to meet me. The local Bayston Hill Library was the next place I visited they were able to tell me of a local village Councillor Ted Clarke. Councillor Ted Clarke was Chair of The Liaison Committee between Tarmac Ltd and the local community; contact was made by telephone and an interview was arranged for the 3rd March 2011. The interview took place at Councillor Ted Clarke's house in Bayston Village, lasting for about one hour. I told Councillor Ted Clarke of my book is and he was very keen to assist, I said that I had contacted Tarmac Ltd. However, Tarmac Ltd said they were not able give me an interview or let me access their archaeological records or the local archaeologist research placed with them. Councillor Ted Clarke told me the that Local Bayston Hill, Parish Council Office had all the records of Tarmac Ltd quarrying applications other important archaeological records; he said that he would be willing to let me borrow all the records appertaining to Tarmac Ltd operations. The interview with Councillor Ten Clarke was very helpful as was the documentation, he also inform me of a Public Right of Way that went straight though Bayston Hill Quarries.

While in Shropshire Archive researching I met a gentleman who was a member of the Clive and Grinshill Conservation Group. He informed me that a man known as Thompson had written five journals on the history on Grinshill Quarries for the Clive and Grinshill Conservation Group. The gentleman said that the journals were now out of print; however he kindly arranged to get the five journals reprinted for me, which I agreed. In a few weeks they were posted to my home address. This is a prime example of just meeting someone through my research; he also put me in contact with a Mr Trevor Kirkham who had worked at Grinshill Quarries for a time, before working for himself. Contact was made with Mr Trevor Kirkham and a date was conveniently made to suit both of us on the 8th November 2011 at Grinshill Quarries. The outcome of this site visit is reported in the appropriate chapter.

Through my research I came across the web-site of The Friends of Llanymynech Limeworks and Heritage Area, which contained a contact list including a Mrs Joan Zoan. I contacted Mrs Joan Zoan and informed of my book is research and asked if she would be able give me an interview; she agreed and took me on a guided tour of the former Llanymynech Limeworks on the 4th April 2013. I also arranged a field trip to the former sites of Lillishall limestone quarries for the 22nd August 2009. After visiting Much Wenlock museum I went on the 14th April 2010 to do a site visit on the former Wenlock Edge quarries. Also in 2009 an Archaeological Record and Heritage Assessment was carried out by Birmingham, Archaeology led by Ric Tyler and Roger White on Lea Quarry Wenlock Edge for The National Trust in which I was involved. I also visited Brown Clee on the10th October 2014. I also contacted Mr Grahame French, Mineral Planning Officer for Shropshire County Council and asked if he might give an interview, which he agreed to on the 16th August 2012. I asked Mrs Harriet Devlin, M.B.E former lecturer at The Ironbridge Institute if she would give me an interview, which she agreed and the meeting took place at The Ironbridge Institute on the 27Th June 2013. Mrs Harriet Devlin had been associated with the former Llanymynech Limestone Quarries; she was instrumental in getting all the different bodies on both side of in England and Wales to work together for the benefit of the former limestone works. Harriet said that there were seven different groups that now owned parts of the former Llanymynech Limestone works and they were all had their own interest at heart, not what was beneficial for the Llanymynech limestone works. The interview with Mrs Harriet Devlin; was excellent it showed what can be achieved when everybody works together. I went to Shropshire County Council offices for the interview with Mr Grahame French. The interview was very constructive on the legislation covering quarrying the earliest to the present day government laws. We talked about all aspects of the mineral quarrying from the environment, conservation and archaeology. I did contact local historian Mr Alf. E. Jenkins who has written books on the Clee Hills and their social and historic background, to ask if he was willing to give an interview, which he declined. Throughout my research most people and organisation had been very accommodation and interested in my research.

2.5.2 Quarrying and transport infrastructure

While the history of quarrying is important, transport of the finished product is essential to the industry. In the earliest times under consideration here, roads were the main means of transport, even if only to get to a river or canal. Albert (1972) writes about the turnpike system of England between the years 1663 to 1840. His first comments are to do with the transport system of the final years of the 17th century, before turning to the creation of the turnpike Acts of Parliament. The writer says that the turnpike trusts were firstly established for a period of twenty-one years, however when the period was about to expire the trustees requested for a renewal of their authority for another term. The initial concept of the turnpike roads was to provide a provisional supplement for parish repairs. However the turnpike trust became more permanent and parish control lapsed; because in the 18th century and early 19th century in England more and more roads came under the turnpike trusts' authority. Albert suggests the turnpike roads were frequently local and when there was a smaller turnpike trust the bigger trust would try and amalgamate them into bigger ones. Initially turnpike roads were administered by a combination of people, but as time went on the management was taken over by people from within the ranks of solicitors, bankers and engineers. He goes on the say that some turnpike trusts were leased out for a fee, which his analysis suggests was very cost effective (1972). Albert writes that canals and railways were allowed to issue shares, but turnpike roads trusts were not allowed too; they raised money from "toll-rents" In the start money came from landowners, which in turn would benefit from the maintained roads. However they raised money through subscriptions and borrowing. Yet with the borrowing became the interest rates on turnpike roads trust, which exposed something of a national market, but he suggests that turnpike trusts were governed by local economic circumstances rather than the rates of interest on the money borrowed. Albert suggests that there was a steady move from regular labour and the trust were looking for a more scientific methods to maintain the turnpike roads. For this the trusts looked to Thomas Telford, who in the 1830s was a Scottish engineer and road builder employed as County Engineer in Shropshire and John. L. McAdam who was a surveyor to the Bristol turnpike trusts in 1816; McAda**m** was credited with inventing macadamisation for road building. There was also an approach to regulate vehicles, i.e. weight, width and the wheels of the vehicles. In the late 17th century justices were appointed to authorize fixed rates for transport of merchandise. From these rates, Albert traces the history of transport charges; their associates with trade volumes and the cost of grain (1972). In this he comes to the conclusion that turn piking had a positive effect. Albert concludes by placing the turnpike system in a much wider context, by examining underdeveloped countries; and at attempts to measure social rates of return on railroads in America. He also notes the return of toll charges on roads to finance and maintain their upkeep. Although it is a good book on the turnpike roads and how they developed, they were usually far away from industrial quarries, such as Brown Clee for example.

Burton, *et al* (2001) writes on the canals of Britain from the early years onwards. With the coming of the canals in the middle of the 18th century a network was set across the country, but canals were limited as where they could be constructed relied on the goodwill of the local land owners. People would invest money in the canal-building enterprise and in return would get issued shares in the canals companies in return for profits. He writes about the construction of the Bridgewater canal, named after the Duke Bridgewater and his engineers. Hadfield (1969 and 1979) addresses the canals of the West Midland and British Canals while Morriss (1991a) writes about the canals of Shropshire as a new way of transporting merchandise, particularly quarried stone. All the authors write how the canals transformed the industries of Shropshire and the country since it was a quicker way to move minerals around, but there were problems because water and slaked lime did not mix. From the books one can see how industries developed around or near to canals, such as quarries, kilns and wharfs. The canals were for a time the backbone of the country because everything could be moved by canal, which connected to rivers and the sea. Morriss' study of the canals of Shropshire (1991a), notes that a number of small canals were built around the late 1760s in the county. The main Shropshire Union canal was authorised in the early 1790s and the main part of the canal was opened in the late 1790s, but the canal remained isolated from the rest of the canal network until the middle of the 1830s. These canals would then link up the main systems across the country, thus providing a connection for the transportation of the quarried minerals from Shropshire quarries and mines. With the building of canals came wharfs for loading and unloading merchandise, canal barges and factories and stables for horses. These authors write about the importance of the development of canals, but the quarry owners were, I suggest, considering ways of improving transport systems too. This can be seen at Lilleshall Stone Quarries who constructed their own canal first.

Robbins (1998) and Wolmar (2007) both write and come to the same conclusion on the coming of the railway age that transformed Britain. There were railways of a description before the 19th century. The tracks were made of stone and iron and they carried wagons from quarries and mines which were pulled by horses. For a time canals and railways existed together, but this would not last and railway companies would try to buy out the canal companies and use the canal system for the railways. The railway companies were owned by shareholders who in return for their investment looked for a good financial reward. In the early years of the 19th century inventor George Stephenson built an eight mile railway in Sunderland; the first to be exclusively machined powered and then in1825 the Stockton and Darlington railway became the first public railway. At first the railways concentrated on passengers and light merchandise. However by the middle of the 19th century railway companies created a countrywide network; this was the beginning of the decline of the canals system that had lasted a hundred years. Morriss (1991b) writes on the railways of Shropshire. Some canals were taken over such as the Shropshire Union canal, and were supported by the railway companies. Wolmar (2007) also writes that before the coming of the railways the only way of traveling was by horse; that is true, however the canals were also faster than horses and they came before railways. The gauge problem was one the major problems in the early years of the railway there need to be a more standardisation. It is clear to see from the literature that the general public and some scientific people were not convinced that the coming of the railway was good. Like any new technical development there were eminent people who considered that the noise from the trains would affect cows and sheep would turn black from the train engine smoke. Wolmar suggests that people were not so enlightened by enterprising ideas that were coming and more to come with the onset of the industrial revolution. The railways would not only enhance the economics of the local regions, such as Shropshire, but also the whole of Britain. With the coming of the industrial revolution the factory owners saw technical innovations advancing their business, however it also worked both ways because the standard of living for the workforces increased. Wolmar writes that the owners on the railways had not expected that passengers would flock to the stations, for a thirty-one mile journey just took two hours by train, undertaken by stagecoach would be twice the time. By the middle of the 19th century approximately 100 million people were using the railways per year on a network of 6,000 miles (Wolmar 2007). With the coming of the railways, Wolmar suggest that the diet of people living in towns and cities greatly improved; fresh milk, meat and fish were now being delivered more quickly. So both the industrialists and the local populations were now benefiting.

So the public could afford days out for the first time they would go the seaside; which for a long time had been the sole activities of the landed gentry, upper class of the country. Wolmar comments that by the 1840s railways were everywhere and trains were going up and down the Britain. In the summer of 1851 the Great Exhibition in London motivated great numbers of people to use the railways. Now the railway had come and there was no going back; quarries and mining industries would have benefited from transporting minerals by railway but also in providing a substantial market for the ballast on which the railway lines actually sat. Wolmar comments that the construction of railway tracks did not cease entirely in the 20th century. New routes were constructed right up to the First World War; however, by the early 20th century competition came from the road transport of goods on lorries, which made it harder for the railway companies to make a profit.

Wolmar summarises that after the boom years of the Victorian period that railways would never be an essential part of our modern lives. The Beeching report published on the 23rd March 1963 recommended the closure of many British Railway lines. However in the 1990s the railways were controversially privatised by the government and to this day the rail companies still get state subsidies. He suggests that with the passing of steam and closure of many branch lines that the railway had lost its appeal. However with the overcrowded roads today that railway journeys could still be the best way to travel. Today in the 21st century quarries and other manufacturing industries are considering returning to rail transport, because it is now more economical such as, the Tarmac Quarries at Sharpstone Hill. The coming of the railway was good for developing industries, but what is not mentioned is the location of quarries in the landscape practically in Shropshire. Some location such as Brown Clee developed their own railways such as the Ditton Light Railway, which connected the quarries to the main line.

2.5.3 Archives and community

To complete this book, research was undertaken in archives in Shropshire, Staffordshire and at the National Library of Wales to access broader historical information and interviews undertaken with people who had interests in individual quarry sites in the county. Archived documents contain reports written, either for planning permission, or on government guidelines. My book is covers six different quarrying sites, thus completing a wider representation of different minerals quarried and providing an overall archaeological and historical picture of the quarrying industry in Shropshire. I acknowledge that there are many more quarries across Shropshire such as, sand and gravel; however I focused on hard-stone quarries of the 19th and 20th centuries. There was only a limited amount of time to research major sites for my PhD and it was decided to concentrate on the more diverse stone quarrying industry.

Early quarrying in Shropshire, before new major technologies were developed, was undertaken by families, or groups of men, with basic tools, working long hours. The quarried stone was transported on the backs of women and single horses from the quarry face, or by pack-horses for longer journeys (Tinder, 2000).

I have also researched and examined legislation covering quarrying in the 19th century to the present day, which has altered considerably from quarries being covered by the Factory Acts of the late 19th to 20th century's legislation "Acts" that now cover all types of quarries in their own right. For example, The Quarry Fencing Act, 1887, The Town and Country Planning Act of 1947, Mines and Quarries Act, 1954, Department for Communities and Local Government, Minerals Planning Guidance 14, 1995 and Department for Communities and Local Government, National Planning Policy Framework, March 2012, p.1).

Adams (2007) identifies the dangers of limestone quarrying and mining. He describes how the Lilleshall Company identified and developed the mineral resources within Lilleshall and Church Aston. Overall, Adams, out of all the individual researchers focuses on more than one quarry or mining site, as can be seen in the chapter on Lilleshall where he provides detailed maps of the Lilleshall Company quarrying and mining activities.

The Friends of Llanymynech Limestone Works gives an ideal example of people coming together for the benefit of the former quarry works. By talking to these people, particularly Mrs Joan Zorn who gave me an excellent tour of the Llanymynech limestone works, it was clear that these people understood the workings of the Drum House, Weighing House, The Hoffmann Kiln, as well as the working lives of the quarrymen and those who stacked the quarried limestone blocks into the kilns for processing. In addition to the workings of the kilns, they clearly understand how the limestone was processed, moved by canal and the associated dangers of working with burnt lime by water. Rhodes' book gives an historical account of Llanymynech and Pant working life, the people involved within the industry that made one of the unique sites in England and Wales (2008). This book recounts the history of the early years of Llanymynech, the technical deployments throughout the year until its closure in 1914. He describes the buildings within the site there, how they were used, the tramways, canals and roads and the uniqueness of having two independent quarries working within close proximity to each other, but being totally independent.

These are two examples where a researcher and a group of people understand the processes associated with Lilleshall and Llanymynech. However, at Wenlock Edge no individuals or groups have been involved in local research. While acknowledging that the National Trust now looks after Wenlock Edge, before this, these former quarries were operated by groups of men who in turn succeeded from the initial purchase of the former monastic lands. Research into the archives documents relating to the quarries and mines in the Shropshire area have enabled me to gain a clear breadth of understanding into this field of operation.

It is clear to see from the research undertaken that geology together with bulk mineral quarrying and mining combine together, as without identifying the source of a given mineral, these processes could not have taken place. The people who purchased the former monastic lands, already knew, I suggest, of the potential of the minerals beneath the ground and saw the opportunity to develop and extract the limestone, sandstone, dolerite and greywacke. From my research, while a clear understanding was gained from reading books, journals, site visits, incorporating guided tours and interviews, the historical information gained from achieved sources proved to be the most valuable. These primary sources such as books, ledgers, notes, and parish records contained first-hand information from people of the 18th, 19th and 20th centuries. They showed the range of processes involved overtime at each site: men quarrying with primitive tools, with horses used for transportation, blacksmith sheds, processing machineries, such as primary crushers, high-intensity screen handling, screen conveyors and grading sheds. The company ledgers from 18th , 19th and 20th centuries shows the amounts of quarried stone needed for detailed works, the cost per ton and how it was transported to each location.

2.6 Archive Research

When researching all the sites one of the first places to visit were the county archives. In the case of Grinshill Quarry, still a working quarry today, Shropshire Archives holds a substantial amount of historical records on the life of the quarry. For instance, a document dated 1722 relates to sandstone from Grinshill Quarry being used in building work carried out for a Mr Jon Dickins by a Mr Ric Brasy (Shropshire Archives, Dickens to

Hill, document no. 112/1/2380, accessed 12/10/2011). Additionally, Carline and Linnel's order books record stone quarried from Grinshill for the years 1771-1772, showing the prices of sandstone, labour, transport cost, together with its eventual use (Shropshire Archives, Carline and Linnel's Order Book, document no. 800 box 20c, accessed 12/10/2011). Information contained in the 1781 Act of Parliament passed Dividing and Enclosing of Common Waste Land in the townships of Grinshill and surrounding parishes, shows the advantage given to landowners and the consequences for local individuals who lost their right to operate small quarries (Shropshire Archives, Dividing and Inclosing Act, document no. 322/9/2, accessed 12/10/2011). Shropshire Archives also contain the order book of Samuel and George Cank for Grinshill Quarry dated 18701873 some hundred years later. This document provides the cost and destination of the quarried sandstone; it also gives terminology used at the time (Shropshire Archives, The Account Book of Samuel and George Cank's business 1870 to 1873, document no. 4756/1/22).

The Clive and Grinshill Conservation Group have published five journals by Thompson about Grinshill Quarries covering the period 1540 until 1923. These journals relate to Grinshill Quarries sandstone being used in the construction and rebuilding of bridges, buildings, roads; also the heyday and decline of this site. These journals also relate the history of the company and its workforce, being a good source of information about the lives of people and their occupations, together with associated trades (Thompson, 2004, 2004, 2005, 2006, 2008 and Historical Directories. www.historicaldirectories.org, accessed 12/10/2011).

The Shropshire Archives also have documents on the people who had owned, leased and worked Grinshill Quarries right up till today. Documents and booklets record the major developers of Grinshill Quarries and, their unique investment in gas powered machinery to cut and dress stone; the precision cutting gave quality and benefit to Grinshill stone; it is also able to withstand atmospheric conditions. This is evidenced in Shrewsbury town where buildings constructed from Grinshill stone can still be seen. The documents also contain pictures of the gas machinery shed, processing stone and cranes lifting quarried stone (Shropshire Archives, Sir Andrew and Acton Reynald, document no. 322/2/428/1, accessed 12/10/2011). The closure of Grinshill Quarries is also recorded together with the re-registering of the quarry under a different name. A ledger of the new company shows the names, shares, occupations, employees and hours worked, as well as invoices for transactions between The Grinshill Stone Quarries Ltd and purchasing contractors (Shropshire Archives, wages ledger for 1926-1929 and other documents no. 6057, accessed 12/10/2011, *Kelly's Directory*, 1922 and Shropshire Archives, invoices for 1923-1937 and other documents no. 6057/2/2 to 6057/2/13, accessed 12/10/2011).

The Shrewsbury Archives and Library and the University of Birmingham Library have been a major source for obtaining books concerning geology, archaeological and historical background in quarrying. Nearly every site researched has been individually affected by the dissolution of the monasteries and records show the progression of ownership of these former Abbey lands (Baker, 2010, Thompson, 2004 and 2006, Blackwell, 1985 and Gough, 1981). The historical directories website gives immense information on the working lives of people, where they lived, where they started work and their chosen trades (Historical Directories. www.historicaldirectories.org, accessed 12/10/2011). Records of the life of The Grinshill Stone Quarries Ltd show how the company changed hands, the financial affairs and the conflicts between the major owners and other directories (Downes, 1980). They also continue to show how Grinshill Stone Quarries Ltd developed under different ownerships and their advancement of new technology such as Bristar power method, a more environmentally friendly way of extracting stone (www.procureit.com, accessed 28/03/2012).

Researching working quarries has been more difficult than redundant quarries, as being able to talk with the proprietors of working quarries has been near impossible, as was the case with Grinshill and Bayston Hill Quarries. However, I was able to conduct personal interviews, firstly, with Mr Trevor Kirkham who gave me a guided tour in and around Grinshill Quarries and told me of the historical background and the future of the site. He also showed me around Corbett Wood, where former quarries site and tracks could be seen (Kirkham, per comm, 08/10/2012). Secondly, Mrs Joan Zoan of The Friends of Llanymynech Heritage Centre gave me an interview and guided tour of this former quarry site (Joan Zorn, Per Comm, 04/04/2013).

Thirdly Mrs Harriet Devlin M.B.E, who lectured at the Ironbridge Institute on Conservation in the Historic Environment, gave me extensive interview about Llanymynech and its quarries, as well as the technological developments made over time at both the English and Welsh Quarries (Harriet Devlin, Per Comm, 27/06/2013). Fourthly, Dr Roger White informed me about the use of limestone at Wroxeter and other Roman sites in Shropshire (Roger White, per comm 04/08/2011). Fifthly, Mr Grahame French, Chief Planner, Shropshire County Council, informed me of the legislation, both old and new, required every time a quarry company makes application for the extraction of minerals. He pointed up the legislation makes specific requirements that the natural environment should be protected before, and after quarrying operations (Grahame French, Pers. Comm. 16/08/2012). I also talked to Councillor Ted Clarke who chaired the liaison committee in Bayston Hill Parish, who was able to inform me about planning applications summited by Tarmac Ltd, Bayston Hill Quarries history (Ted Clarke, per comm, 03/03/2011).

As with all the quarry sites researched, whether worked or redundant, legislation still applies in some form or another. Achieves, books and government web-sites provide enormous information on Acts of Parliaments from the middle of the 19th century to the present day. There were no safety regulations to cover mines yet alone the quarrying industry until a major accident in Husker Colliery, near Silkstone, in 1838. The Metalliaferous Mines and Quarries Act 1842 was subsequently instigated (Mills, 2010). Additionally, the government altered the 1891 Factory Act, so that it covered open air works such as quarrying. It was 56 years later, in 1947, before the first planning acts would come into force. Since then legislation has greatly changed, as today the laws covering the quarrying industry are very stringent, covering the environment, safety, aftercare and archaeological finds, are just a few examples (Mill, 2010 and Grahame French, per comm. 16/08/2012).

2.7 Law and Legislation, concerning The Quarrying Industries in the 19th and 20th Centuries

2.7.1 Introduction

This section presents the appropriate laws and legislation that have and still cover the quarrying industries today. These laws cover areas which include: planning, environment, conditions of work safety, flora & fauna, geology, mineral extraction and archaeology. This chapter will examine the laws of the 19th century to see how they have changed through time to the present day. I will be examining The Quarry Fencing Act, 1887 and The Quarries Act, 1894 to see how early Acts of Parliament covered and protected the workforce and their employers in the quarrying industries. Additionally examination will be made to determine how the general public was protected from the hazards of the quarrying industry. I will also investigate 19th century environmental issues associated with the quarrying industry, to see how those issues were addressed then and compare how they are addressed today.

Health and Safety steadily became an area of concern in association with the mining industry during the time of this research period (Brown, 2005).

Although safety regulations did exist in the 19th century, there were no Planning Acts in existence until 1947. The quarries that were extracting minerals, such as limestone, were considered by the law to be the same as those extracting coal. At this time 'quarrying' could also be known as 'mining'. At the same time an open cast working area was known as a 'quarry', while underground workings were called 'mines'. It was argued in the case of *Kent v. Astley,* that as quarries were open to the air they should not be considered as a 'factory', however the law was altered to cover open air premises so that the Factory Act 1891 could be put into practice. The only legislation in existence for quarries before 1947 was brought about in response to the increasing amount of accidents. This legislation covered health and safety issues in mines (Mills, 2010 and Grahame French, Pers. Comm.16/08/2012). The 1830s saw the labour force raising their concerns about Health and Safety in the workplace "mines"; however it took some eight years before the Miners' Association was established. Few records have been found, so it is not possible to estimate the full amount of accidents that were happening in the mining industry at that time.

Throughout the 1840s The Miners' Association tried to highlight safety issues in mining. To enhance their concerns they published songs and poetry such as, "The Miners' Grievances" and "The Colliers Rant". These were published and sold with three aims; to raise funds, promote awareness of working conditions in mines and solicit public support for their campaigns (Mills, 2010).

The conditions of work in mines came to the forefront in 1838 after an accident at Husker Colliery, near Silkstone. This accident made the public more aware of the conditions of the miners' working environment. A powerful freak thunder storm caused a stream to flood, which flowed into the ventilation shaft. This then flooded the mine causing the death of 26 children; 15 boys, aged from 7 to 16 years and 11 girls aged from 8 to 17 years old. This tragedy came to the notice of Queen Victoria, who ordered an inquiry (Gallop, 2010 and Mills, 2010). The Royal Commission, chaired by Lord Ashley, investigated working conditions in the mining industry; in 1851 he inherited the title of Lord Shaftesbury. The Royal Commissioners visited different collieries and mines gathering information from personal accounts of mineworkers about their working conditions; this was often against the mine owners' wishes. The Commission's findings were published in May 1842 conveying the dreadful conditions that had been found and that, significantly, the highly dangerous and adverse working conditions were the norm. These made accidents more likely and revealed the brutality of long working hours experienced by women and children, as well as the danger of silicosis, more often suffered by the miners themselves. Women and children worked as much as 12 hours a day, receiving less wages than the men. Children as young as six years old worked as trappers, opening and shutting ventilation doors into underground mines. The Royal Commission's report resulted in a public uproar about the working conditions of miners, equally expressing concern about the ages of the children who worked down the mines. The report combined with the public outcry resulted in the 1842 Mines Act being brought into force (Mills, 2010 and Grahame French, Pers. Comm. 16/08/2012).

The first inspector of mines to be appointed under the 1842 Act was Hugh Seymour Tremenheere, in 1843. He only had limited powers under the Act, but took up many prosecutions. Tremenheere stressed the need for education of the officers of the mines and the working men, also the need for education in the community, which would help towards enabling the provision of a safer working environment. On 22nd July 1850, Lord Carlisle introduced a bill in parliament to regulate the coal mining sector. He called on The Home Office to appoint impartial inspectors empowering them, with limited authority, to enter and examine any mining premises. On 14th August 1850 the Bill received Royal assent after a few concessions (Mills and Grahame French, Pers. Comm. 16/08/2012). Reforms seemed to be gathering pace now because in 1862 the government of the day appointed a Royal Commission to investigate working conditions in Metalliferous Mining Sector "quarries". This led to The Metalliaferous Mines Regulations Act of 1872, which is discussed later (Mills, 2010). Successive governments started and continued to provide guidance on all aspects of mineral planning to local governments, called "The Mineral Planning Authorities document" which also provided guidance to quarry owners and operators.

2.7.2 Quarry Acts

The Quarry Fencing Act 1887, which came into force on January 1st 1888, covered quarries that extracted: stone, slate, lime, chalk, clay, gravel, or sand. Quarries were considered dangerous to the general public if they were in open land, or within 45.72 metres of a highway, or dedicated public access route, which might not have been securely and sufficiently fenced. The quarry owners were therefore deemed responsible for providing reasonable fencing for the prevention of accidents. If the fencing was judged not to be adequate, the quarry owners were considered to be causing a nuisance and liable to be prosecuted under The Public Health Act 1875 (The Public General Acts of the United Kingdom, 1886).

Between 1878 and 1894 'surface quarries' were regulated only by The Factory Acts as far as it applied. This Act was not applicable to underground quarries, referred to as 'mines', which were therefore regulated by the Metalliferous Mines Regulation Act, 1872. So at that time there were two different regulatory bodies. The 1878 Factory and Workshop Act, stated that no premises were to be excluded from this Act on the grounds that they were operated in the open air; this countered arguments made at the time that quarries were worked in the open air and so were not deemed to be 'factories'. There was some argument about the terminology used, in that 'surface quarries' were not deemed to be 'mines' and vice versa. Because of this dispute about the accepted terminology, quarry owners considered that the 1878 law was not applicable to their quarries (The Public General Acts of the United Kingdom, 1878). This was brought to a close by the 1891 Factory and Workshop Act which included these surface quarries under its "special regulations" which were adapted to meet the unique dangers and risks associated with operations in surface quarries. In December 1892 the Secretary of State designated these quarries as 'dangerous places of work' and therefore subject to special rules. In 1894 the Quarries Act was introduced. This extended the powers of the Metalliferous Mines Regulations Act 1872, giving the inspectors the authority to enforce provisions for notification of any accidents and the undertaking of prosecutions, which in time led to the establishment of The Quarry Inspectorate (Mills, 2010. pp. 39-69 and United Kingdom and Quarries Regulations in 1910, accessed 27/09/2012).

In 1893 the Home Office started an inquiry under the chairmanship of Frank Wardell, the Inspector of Mines, and this brought about the regulation of all open quarries that were more than 6.1 metres in depth, which led to the establishment of the 1894 Quarries Act. This Act of Parliament mainly utilised Parts 1 and 3 of the Metalliferous Mining Regulations. Although it stipulated no universal code of working practice, it did contain provision for the implementation of "special rules" (Mills, 2010). The 1894 Quarries Act brought every surface quarry, which was more than 6.1 metres deep, under the Metalliferous Mines Acts of 1872 and 1875, so that they therefore fell under the authority of the inspectors appointed under the said Acts. The Inspectorate made many contributions to improving the occupational health of the metalliferous miners. Following the encouragement of the Inspectorate, a major improvement in

the early 1890s was the establishment of voluntary first aid and an ambulance station at the working mines and quarries. By 1899 the Inspectors of Mines had reported that over 2,000 quarries had established special safety rules. However, in 1905 these inspectors still found that accidents brought about by dangerous blasting practices, were caused by negligence in the observation of the special regulations laid down under the 1894 Quarries Act (Mills, 2010 pp. 127-151).

In the early 20th century The Metalliaferous Mines Regulations Act was subject to a lengthy enquiry, brought about by further developments in colliery regulations. A Royal Commission was set up to examine general safety measures, as well as what methods were in place for the prevention and control of dust from explosions. The findings of this Commission caused a re-examination of The Metalliaferous Mines and Quarries Act 1842 to be instigated (Mills, 2010). The first Royal Commission instituted that a second Commission, to be carried out in greater depth, should be enacted, which took place in 1910, under the chairmanship of Henry Cunynghame, the Undersecretary of State. This second Commission of enquiry served to uncover numerous weaknesses in the regulations, which highlighted the difficulties and concerns first raised by the Mining Inspectorate, because of the in-exactitude of the wording of the original statues in The Metalliaferous Mines Regulations Act of 1872 (Mills, 2010, pp.151-183).

The Royal Commission of 1910 published its report over the years 1912 and 1914, which made extensive recommendations. They suggested that additional Inspectors should be appointed and there was an urgent need for the consolidation and simplification of the law. Sections in the Act relating to supervision and management were to be brought into line with the current Coal Mines Act. It was also suggested that the responsibilities and roles of mine and quarry owners and their agents ought to be more clearly defined. The commissioners also strongly suggested that every mine and quarry employing more than 12 men underground must be under the control of a registered manager. These managers should conduct a daily examination of the working environment and record their findings. The Commission also strongly recommended that strict safety regulations should be monitored in relationship to the level and fall of the ground at exit and egress. Stipulations were put into place regarding the safe use of explosives, haulage, drainage of water and the maintenance of standards of ventilation and air temperature (Mills, 2010, pp.39-69).

The 1910 Mine Accident (Rescue and Aid) Act was also to be revised. It was suggested that all metalliferous mines and quarries employing more than 10 men, should be required to have for immediate use a first aid kit and stretcher. In addition, quarries employing more than 100 men must have an ambulance on site and persons trained in first aid at all times (Mills, 2010). The Acts of Parliament that already existed covering mining and quarrying in the late 19th to early 20th century seem not to have been updated until after 1947. Extensive research cannot find any addition to these laws. However from 1932 onwards mineral workings were subject to limited planning control, certain aspects of which referred to safety and conditions. With the commencement of World War II, under wartime emergency powers in 1943-1948, the government imposed an "Interim Development Order Consent" on quarrying. The Town and Country Planning Act of 1947 reviewed existing workings and made planning permission compulsory for new quarries (Grahame French, Pers. Comm. 16/08/2012 and Minerals UK, www.bgs.ac.uk/mineralsuk/planning. legislation, accessed 19/10/2012).

Legislation governing mining and quarrying has developed over the years, particularly prompted following major disasters. However, early legislation only covered the "Health and Safety" aspects of working conditions in quarries or mines, but did not cover the environmental issues caused by quarrying and mining. All the lands containing mineral rights were privately owned so that there were few regulations governing the immediate environment around the quarry sites (Mills, 2010, Grahame French, Pers. Comm. 16/08/2012 and Minerals UK, www.bgs.ac.uk/ mineralsuk/planning.legislation, accessed 9/10/2012).

2.7.3 Town and Country Planning Act 1947

The Town and Country Planning Act of 1947 instigated a new planning system which covered the ordinance of quarries and mines. The objective of the new Act was to replace all existing systems of planning control. It also prohibited the further carrying out of any kind of development plan whatsoever before, or after, the coming into operation of this Act of Parliament. This new Act of Parliament had to deal with quarries that were already in existence prior to 1947. Many quarries in Shropshire fell into this category and they were required to apply for an Interim Development Order Consent (IDO). Quarry owners and operators who wished to continue operating lawfully after the establishment of the 1947 Act of Parliament had to register an IDO with their local authority; then the quarries were allowed to continue operating within the structure of the law

(Grahame French, Pers. Comm. 16/08/2012 and Heap, 1947). However this brought significant problems because some planning authorities had very little technical background or any historical understanding of the particularities of quarrying and the quarrying industry. Not understanding the proper procedure, these authorities were granting IDO's and so, in effect creating large areas of land with lines drawn around them where no proper control could be exercised. The IDOs were being issued, but no clear instruction to the land owners, quarry owners and operators were being given under these orders. No instructions as to what quarrying could take place were given; the only injunction put into place was that the land should be reinstated once quarrying had ceased (Grahame French, Pers. Comm. 16/08/2012 and Heap, 1947). Since the coming of the 1947 Act there have been a series of developments in planning policies, including those concerning mineral waste. This situation continued into the 1980s and early 1990s, with IDO consents were being issued and used without any proper planning control, so that the government of the day had to readdress the situation. This resulted in the government publishing Mineral Planning Guidance Notes 14 (MPG14) (Grahame French, Per. Comm. 16/08/2012 and Heap, 1947).

2.7.4 Mineral Working Act 1951

In the 1950s the government started to address the management and control of quarries and mines. The 1951 Act of Parliament was set up after the Kennet Committee report of 1938 and the Waters Committee report of 1946. This followed serious attempts to control the environmental concerns of mining, because of the ironstone workings in the East Midlands. These workings were considered to be the first major opencast workings in Britain. A restoration fund was set up, which was funded by mineral producers and mineral rights owners who contributed to a per-ton fee. This was then used to pay for the restoration of the landscape after quarrying had ceased (Minerals UK, www.bgs.ac.uk/mineralsuk/planning .legislation, accessed 19/10/2012).

2.7.5 The Mines and Quarries Act 1954

The Mines and Quarries Act, 1954 went further than the previously discussed Quarry Fencing Act 1887. It covered the safety of the general public and would make new provisions for the management of mines and quarries. Also the Act required the proper management of the safety, health and welfare of persons employed in the mines and quarries. It regulated the employment of women and young person in mines and quarries. In addition the Act required the proper fencing of disused mines and quarries, thus providing secure fencing for the safety of the general public who might walk near the perimeter of the quarries or mines (Mines and Quarries Act, 1954, www.legislation.gov.uk, accessed 29/01/2012).

2.7.6 Mines and Quarries (Tips) Act 1969

The 1950s and 1960s saw concerns being raised about the effects that quarries and mining were having on the environment and landscape; especially the dangers of disused quarries and mine being used as mineral tips. The Aberfan disaster in 1966, which was caused by an old waste-tip from a mine, could have in some way influenced the government to pass The Mines and Quarries Act, 1969. This Act of Parliament was an extension of The Mines and Quarries Act, 1954, which did not mention mineral waste-tips precisely in its provisions. There was only reference to making quarries and mines safe by erecting fencing adjacent to the perimeter, so that the public would not fall into the quarries (Mines and Quarries (Tips) Act, 1969, www.legislation.gov.uk, accessed 29/10/2012).

2.7.7 The Mines and Quarries (Tips) Regulations, 1971 and The Quarries Regulations 1999

The details for compliance with the 1969 Act are laid out in The Mines and Quarries (Tips) Regulations, 1971. Subsequently The Quarries Regulations 1999 required that the mineral waste-tips must be designed, constructed, operated and maintained so that any instability or movement which is likely to cause a danger to the health and safety of any person is avoided. It also specifies that geotechnical and other measures should be taken to ensure the tips do not move. Additional legislation that could have some bearing on the design, construction, operation and disposal of mineral waste-tips was the (Rivers Prevention of Pollution Acts and the Clean Air Acts of Parliaments Minerals UK, 2001,www.bgs.ac.uk/ mineralsuk/planning.legislation, accessed 19/10/2012, The Mines and Quarries (Tips) Regulations, 1971 and The Quarries Regulation, 1999 www.legislation.gov.uk, accessed 29/10/2012).

Today the Department for Communities and Local Government (DCLG) provides guidance on the legal framework concerning the health and safety of employees, local residents and the general public. There is also a planning framework, which concerns the environment. Under the DCLG there are guidance notes on most forms of land development in

Britain, which includes quarries and mines. This now leads to the Mineral Planning Guidance Notes 14 (MPGs14), which come under the DCLG www.communities.gov.uk, accessed 12/10/2012).

2.7.8 Mineral Planning Guidance Note 14

As discussed before, in the late 1980s to early 1990s there were many IDOs where consent was given, but without proper planning controls being required. So the government of the day published Mineral Planning Guidance Notes 14, 1995, (MPG14) and with it went a lot of other documents. This new policy would bring together all the local authorities and stop inconsistencies in the implementation of mineral policies in different areas. By initiating this new policy the government provided centralised guidance on mineral planning in the country. With the coming of MPG 14, it was made compulsory for quarry and mine operators who already had consents under IDOs, to re-register for new operating agreements. They were now re-registered under modern working conditions; this was the first review deadline to be met. However the government of the day did not stop there; they required that there should be a periodical review every 15 years after the original registration (Grahame French, Per. Comm. 16/08/2012 and Department for Communities and Local Government, Minerals Planning Guidance 14, 1995, www.comunities.gov.uk, accessed 30/07/2012).

The environmental impact of the quarrying industry raised significant questions. When re-registering, or reviewing, previous mineral workings, all of the customary environmental tests that had been applied originally were voided, and the industry had to make new application to obtain quarrying permission, just as if it were the very first application to be made. The quarry owner or operator, before submitting a periodical review, has to provide an "environmental impact statement" which must contain comprehensive information about the likely key environmental effects of any proposal. This was onerous to both the planning authorities and the mineral industry, because the quarrying industry could suggest that they already had established principal mineral workings (Grahame French, Pers. Comm. 16/08/2012, DCLG, Minerals Policy Statement 1, 2006 and 2, 2005, www.comunities.gov.uk, accessed 15/05/2012, DCLG, Minerals Planning Guidance 14, 1995, www.comunities.gov.uk, accessed 30/07/2012, DCLG, Planning and Mineral: Practice Guide, 2006 and DCLG, Environmental Impact Assessment and Reviews of Mineral Planning Permission, 2008, www.comunities.gov.uk, accessed 31/01/2012). When submitting for a periodical review, quarry owners or operators have to take into account ecological considerations especially for Site of Special Scientific Interest (SSSI). In the Mineral Policy Statement 1, there are guidance notes for potential quarry workings or quarries applying for a periodical review; in particular, guidance notes on the protection of the countryside, heritage and wildlife (DCLG, Minerals Policy Statement 1, 2006, www.comunities.gov.uk, accessed 24/08/2012). The mineral industry as a whole has considered the periodical review as onerous and is calling for something with a lighter touch, when re-registering (Grahame French, Pers. Comm. 16/08/2012).

There is now a greater focus and concern for the protection of the environment and what possibilities may exist to mitigate against the environmental impact of mineral workings on the environment. The MPGs were a fundamental way of doing that, but then the technical guidance notes were replaced by the Mineral Policy

Statements 1 and 2 in the middle of the first decade of the 21st century. With the MPGs and MPSs being in place now, these have proved to be a shorter and more concise statement of national policies concerning mineral planning. These were guidance notes, which have been separated out into technical advisory notes for working quarries and for the control of dust and environmental amenities. Even though these proved to be of help to the mineral industry, the "National Planning Policy Framework" (NPPF), published in 2012, would again simplify and streamline all policies relating to all aspects concerning the quarrying industry. The (NPPF) is sixty-five pages from beginning to end, not hundreds of documents, pages and pages long (Grahame French, Pers. Comm. 16/08/2012, DCLG, Minerals Policy Statement 1, 2006 and 2, 2005, www.comunities.gov.uk, accessed 15/05/2012 and DCLG, National planning Policy Framework, March 2012, www.comunities.gov.uk, accessed 24/08/2012, p.58).

2.7.9 National Planning Policy Framework

The National Planning Policy Framework is an important change to mineral planning policies and how they are interpreted. A lot of important information is contained in this new policy. However the parts of the policy concerning mineral planning are

sections on "Conserving and Enhancing the Natural Environment" and "Conserving and Enhancing the Historic Environment". There are technical guidance notes associated with the NPPF, which indicate that there is a requirement to refer to these guidance notes for more details on issues such as, the dust and noise related to quarrying (Grahame French,

Pers. Comm. 16/08/2012 National Planning Policy Framework, March 2012, www.comunities.gov.uk, accessed 24/08/2012, pp. 25-29 & 30-32).

It is also important to take into consideration that minerals are essential to supporting and maintaining economic growth and our quality of life. With the growing concern for the maintenance of the environment, it is also important that there is an adequate supply of minerals such as quarried stone for the maintenance of historic buildings throughout the country. However, it must be understood that minerals are a limited natural resource and can only be quarried at specific sites. Therefore it is very important to make the greatest use of these minerals (DCLG, National Planning Policy Framework, March 2012, www.comunities.gov.uk, accessed 24/08/2012, pp.32-36).

The natural environment is a major issue when considering a mineral planning application. Today, the planning system should contribute to the enhancement of the natural environment, whether nationally or locally. Consideration should be given to the protection and conservation of any geology, valuable landscape and the wider benefits of the ecosystem. The government's policy is to stop the decline of our biodiversity, so any planning proposal should contain the minimum impact on the overall biodiversity. Today, local mineral planning authorities have to take into account the general public's concerns and issues on the local environment. Therefore local planning authorities strive to prevent new and existing developments from contributing to the possible excessive effects of unacceptable levels of pollution from noise, and that created in water, air and soil from mineral waste. These points are not only for the safety of the public and the environment, but also for the quarrying industry. The overall aim is then to minimise pollution and other adverse effects on the local and natural environment (National Planning Policy Framework, March 2012, www.comunities.gov.uk, accessed 24/08/2012, pp.25-27).

The NPPF contains many detailed guidance notes and principles that should be followed by local planning authorities and mineral developers. It points out why planning permission should be refused if mineral developments cannot avoid significant harm to the biodiversity. Taking this into consideration an alternative site with less harmful bearing on the environment should be found, especially when considering planning applications which might approach on land within, or outside, a SSSI. In addition, planning authorities should aim to conserve and enhance these sites. Thus developments on SSSI's such as, "quarrying" which have adverse effects to the site should not be ordinarily permitted. There is an exception, but only where the benefits of the developments on a SSSI clearly outweigh the impact to the site (DCLG, National Planning Policy

Framework, March 2012, www.comunities.gov.uk, accessed 24/08/2012, P28). However, development proposals that contain objectives that will conserve and enhance biodiversity should be permitted; this sort of proposal should be encouraged. There is also guidance on why planning permission should be refused; this is the case if the development results in the loss or deterioration of irreplaceable habitats. This includes the loss of ancient woodlands as well as individual mature or historic trees, growing either in, or outside of ancient woodlands. Again, development should consider if the benefits clearly outweigh the loss of these habitats (DCLG, National Planning Policy Framework, March 2012, www.comunities.gov.uk, accessed 24/08/2012, p.28).

The NPPF policy also suggests that wildlife sites in England should be given the same protection as European sites. These potential wildlife sites could become Special Protected Areas within Special Areas of Conservation. Existent and proposed Ramsar sites are wetlands which are defined as marsh, peat land, fen and water areas, whether natural or artificial (Natural England, www.naturalengland.org.uk, accessed 26/10/2012, p.28). These sites may have stagnant and flowing water which may be fresh, or include areas of brackish salt marine waters. It does not matter whether these sites are permanent or temporary. There are many more areas that fall under "Ramsar Sites" such as, banks of a stream, rivers and ponds (Natural England, www.naturalengland.org.uk, accessed 26/10/2012 and (DCLG, National Planning Policy Framework, March 2012, www.comunities.gov.uk, accessed 24/08/2012, pp.28-29).

Another major part of the NPPF is the section that deals with conserving and enhancing the historic environment. It provides guidance for local planning authorities as to how to set out a positive approach concerning the conservation and preservation of the historical environment, so that it can be enjoyed by all. This includes those historic heritage assets which are deemed to be most at risk (DCLG, National Planning Policy Framework, March 2012, www.comunities.gov.uk, accessed 24/08/2012, pp.25-29).

2.7.10 The Sustainable use of Minerals

This part of the NPPF is probably the most important section as it concerns the sustainable use of minerals. The significance of minerals in the sustainability of our country's economy has been preeminent for centuries. It has always been understood that there needs to be a sufficient supply of minerals for the maintenance of our historical buildings, roads and industry. However, minerals are a limited resource and can only be quarried where they are found, so it is imperative that minerals are put to the best economical use, which then safeguards them for the future (DCLG, National Planning Policy Framework, March 2012, www.comunities.gov.uk, accessed 24/08/2012, pp.32-36).

The NPPF guidance to local authorities on preparation of local plans for developers should, where it is possible, take into consideration that recycled stone and mineral waste might be used in preference to newly cut stone. Site visits showed this is to be most obviously seen at Grinshill Stone Quarry, near Shrewsbury and Lea Quarry, Wenlock Edge. Grinshill Stone Quarry is still a working quarry and one can understand the government's policies on the use of quarry waste stone, particularly where the possibilities of extension to the quarry are limited, or not at all possible. In the case of Lea Quarry, now dormant, this has possible reserves of stone not yet quarried which could be used in the future maintenance of local and national heritage. This is a prime example of fully utilising the site before considering new extractions or extensions to existing quarries, therefore conserving local and national resources and the environment (DCLG, National Planning Policy Framework, March 2012, www.comunities.gov.uk, accessed 24/08/2012, pp.33 &14).

It is of great importance that special mineral resources are safeguarded and clearly defined locally and nationally, so appropriate policies can be put into place. Therefore, by knowing where these special minerals sites are and what they contain, it will be possible to stop the destruction (sterilisation) by other industrial developments. By defining and recording these areas it is not implied that all will be worked. Mineral Consultation Areas will be set up in order to safeguard these minerals. The benefit of this is the re-use of existing or creation of newly planned rail heads, to link quarries to the national railway. This is the case with Bayston Hill Quarries, which have a disused rail siding and are in the process of re-developing the rail sidings so that they can transport greywacke stone by railway (DCLG, National Planning Policy Framework, March 2012, www.comunities.gov.uk, accessed 24/08/2012, p.33).

As already discussed, in the 21st century the environment became a major concern, so it was necessary for local planning authorities to set out their environmental criteria with guidance from the policies of the NPPF. Planning applications will be assessed against this framework. This will ensure that any permitted mineral operations will not have unacceptable adverse effects on the environment by taking into consideration the impact that quarrying operations could have on the natural and historical environment. This would include the adverse effects of noise, dust, visual intrusion, traffic i.e. lorries, all of which can be detrimental to human health. It will also take into the equation the effects of the possible instability of quarry wastes, the slopes formed by quarrying, and their need for stabilising measures. The aftercare of disused quarries, once having been backfilled to obviate the dangers of subsidence caused by quarrying and mining, should also be monitored. This was not done at Lilleshall Limestone workings where an old limestone mine collapsed. Work was subsequently carried out by Messrs Wrekin Construction in 1997 to fill in the former mine with 100,640 tonnes of bulk infill and 46,540 tonnes of cement pressure grout, which totalled 147,160 tonnes. This gives an idea of how much limestone had been removed during the past two centuries (DCLG, National Planning Policy Framework, March 2012, www. comunities.gov.uk, accessed 24/08/2012, p.33).

Restoration and aftercare is essential and should be considered when giving planning permission. It is fundamental that policies are put into place to ensure that land being used for quarrying is reclaimed at the earliest possible opportunity, so that it can be returned to agricultural use, or to the natural environment, in which case such sites can be replanted with trees for the benefit of the local population (DCLG, National Planning Policy Framework, March 2012, www.comunities.gov.uk, accessed 24/08/2012, p.33).

The NPPF emphasises that more co-operation should take place between all planning authorities across the country. This would assist in the formation of better planning policies for the more effective use of minerals within the construction industry and manufacturing processes. Again there is a focus on having enough minerals in reserve, such as stockpiling important minerals for conservation and restoration work. The point being made in this document is that better care should be taken of the management of quarry sites, whilst at the same time making sure that the quarries we do have are used economically, so as to preserve them for the future. In addition, there is a need to work together with developers and other planning authorities across the country as a whole (DCLG, National Planning Policy Framework, March 2012, www.comunities.gov.uk, accessed 24/08/2012, pp.34-36).

2.8 Archaeology and Mineral Extraction

English Heritage (now Historic England) also provides guidance notes on mineral extraction and archaeology. As already discussed, minerals are essential to the nation's economy (Mineral Extraction and Archaeology: A Practice Guide, 2006, www.english-heritage.org.uk, accessed 11/07/2012). This document acknowledges both economic and archaeological values.

They point out that archaeological remains are irreplaceable, are a limited resource. In the majority of cases these sites are highly fragile; which makes them vulnerable to damage, sometimes beyond repair. As with minerals, all are not as valuable as each other, however in the case of archaeology, some resources are of both national and international importance. The hillfort on Brown Clee was irrevocably damaged by quarrying in the mid-1800s which removed part of the ramparts, before the appropriate legislation had been brought in (Shropshire Archives, document no, XLS7956, The Hill Forts of Shropshire, accessed 22/12/12). Such sites need the maximum protection from development such as quarrying, which may be in competition for the same site. So when archaeology is discovered the only way of finding its value is through archaeological investigation. As discussed before, it is the role of the mineral planning authorities to recognise the needs of both the developers and the historical and archaeological environments (Mineral Extraction and Archaeology: A Practice Guide, 2006, www.english-heritage.org.uk, accessed, 11/07/2012).

In 2009 an archaeological record and heritage assessment was carried out at the former Lea Quarries, Much Wenlock, Shropshire. This is a site of extreme uniqueness to the archaeological and historical background of the quarrying industry in Shropshire, because it was only mothballed in 2007, so this makes it one of the last working quarries in Much Wenlock. The Lea Quarries assessment recorded and photographed all the limestone processing buildings. It was considered that part of the former site, for example Lea Quarry North, had a high evidential value in its modern quarry extraction and processing methods of the limestone, however, the limestone is considered to be of low value. The economic cost of extraction and processing of the limestone was no longer viable at the site. It was considered that an alternative use for this site might be found, however the machineries would be difficult and expensive to save for heritage purposes. Today the former Lea Quarries are used as storage for large lap fencing (Birmingham Archaeology, 2009). I suggest that an industrial museum would have been a better alternative for the site. Therefore preserving the former Lea Quarries buildings and limestone processing machineries would provide an insight of the past within a living museum. Thus giving future generations an idea of the archaeological and historical significance of the quarrying industry in 19th and 20th century Shropshire.

2.9 Discussion

This particular section has demonstrated how the law has been developed from the 19th century to the present day. One can see the need for past and present governments to institute Acts of Parliaments to protect the workforce, particularly in the 19th century. After the mine flood disaster of 1838, which killed 26 children who had been working underground, it was quite apparent that there was a need for the government to act. The 7th Earl of Shaftesbury, Lord Ashley was fundamental in bringing changes to the mining industry and factory workers. It was apparent that there were no restrictions on the actions of the mine owners who were having it 'all their own way' and the safety of the workforce was not top of their agenda. In addition to the provisions of this Act for safety of the workforce, the owners were expected to consider public safety around the perimeter of their mines and quarries. Quarries before 1894 did not have their own identity, within the law, and fell under the Factory Act, and Factory and Workshop Acts. On the grounds that quarry workings were mostly open spaces, the owners stated that the Factory Acts were not appropriate.

The Metalliaferous Mines Regulations Act, 1872, can be considered as the first step towards safety regulation; being subject to two enquiries, it brought about the introduction of special regulations in mining and quarrying. Having inspectors who had some powers when the law was first introduced and, as time went on, their powers increased to cover the inspection of mines and quarries and the ability to bring about prosecutions when necessary. Legislation concerning quarrying and mining in the late 19th and early 20th centuries was at an early stage in its development. The coming of World War II saw more positive regulation becoming law. In 1947 the Town and Country Planning Act came into force which initiated some concrete planning policies for Shropshire and other local mineral planning authorities across the country.

However, the late 1980s and the early 1990s saw the MPGs, and particularly MPG14 provide more in-depth guidance policies for mineral planning authorities and developers. Together with the Mineral Policy Statement 1, these policies have brought about some structure to mineral planning across Shropshire and the country. Though these policies did help both mineral planning authorities and development planners, they were very large documents. So when the developers and mineral planning authorities had to summit periodical reviews every 15 years, it took a tremendous amount of time to comply with this legislation.

Then the NPPF was introduced which streamlined the mineral planning process for the Shropshire Mineral Planning Authorities and all other Planning Authorities across the country. These guidance notes covered all aspects of planning, especially that concerning mineral developments. One of the major points concerning mineral planning and existing quarries is the environment and the effect quarrying has on it. The environment consists of many factors: SSSIs, ecosystems and the noise associated with blasting; this was a major concern at Knowles Quarry, Wenlock Edge, in the early 1900s. Today local residents in Shropshire are concerned about the amount of traffic in and out of quarries. Another concern is what happens to quarry sites after quarrying operations have ceased; what restoration will take place and its aftercare? We must also remember that archaeological discoveries are sometimes found when new quarrying sites are opened up. These need to be properly examined and recorded and are part of our heritage today and for the future (Ancient Monuments and Archaeological Areas Act 1979, accessed 20/07/2012).

2.10 Conclusion

This chapter has been illuminating and key to the research into 19th and 20th century quarrying in Shropshire. This research has shown the development over time of mineral policies and laws that regulated production in the Shropshire quarries which produce limestone, greywacke and sandstone. Research has shown how health and safety of the workforce was addressed in the 19th century, together with laws governing the employment of children, women and men. The 20th century saw the development of mineral planning laws and guidance notes for mineral extraction in Shropshire and the country as a whole. It is of interest to note that some of the environmental concerns raised today were already important for people in the late 19th and early 20th centuries. These issues will continue to affect mineral quarrying now and into the future, so essentially the ways of managing the environment together with the economic needs for quarrying, must be balanced.

CHAPTER 3

GRINSHILL QUARRIES, CLIVE

3.1 Introduction

This chapter will research an active sandstone quarry worked by Grinshill Stone Quarries Limited. Grinshill village and quarry lies approximately 7 miles NE from Shrewsbury on the A 49. The sandstone quarried at Grinshill is of a high quality, suitable for both new buildings and restoration work. The sandstone is available in a creamy white and light reddish colour being used extensively throughout the United Kingdom; the Grinshill stone was the principal building stone in the county from the 16th century.

3.2 Geology of the Grinshill area: SJ 526238

The geology of the Grinshill area is mainly comprised of sandstone and flagstone deposits from the Triassic period from approximately 240 million years ago. The bedding is inclined at an angle of approximately 5 degrees northwards into the North Shropshire syncline and towards the Cheshire Basin (Toghill, 2006 and Greenwood *et al,* 1913). On top of this sandstone is the Waterstone or Tarporley Siltstone Formation; this is only found in the Grinshill and East Baschurch areas. The name Waterstone almost certainly derives from its watered-silk like appearance; this is produced by the sheets of white mica, a muscovite mineral, which makes up the stone (Toghill, 2006). The Upper Keuper Formation, now known as the Mercia Mudstone Formation, is made up of upper, middle and lower layers, of which the Lower Keuper sandstone is found in the Grinshill Quarries. The Lower Keuper sandstone consists of fine grained, white and reddish sandstone and is most desirable as a building stone; however the lower beds are harder to quarry. The sandstone is an attractive cream colour and because of the fineness of its grain, it is very durable. In Shropshire there was igneous activity, which can be seen around Grinshill, particularly in Grinshill Quarries where it formed the Dolerite Dykes (Toghill, 2006, Ward, 1841 and Murchison, 1839). Near to Grinshill Quarries is the Brockhurst Fault which runs to the east of the quarries; this

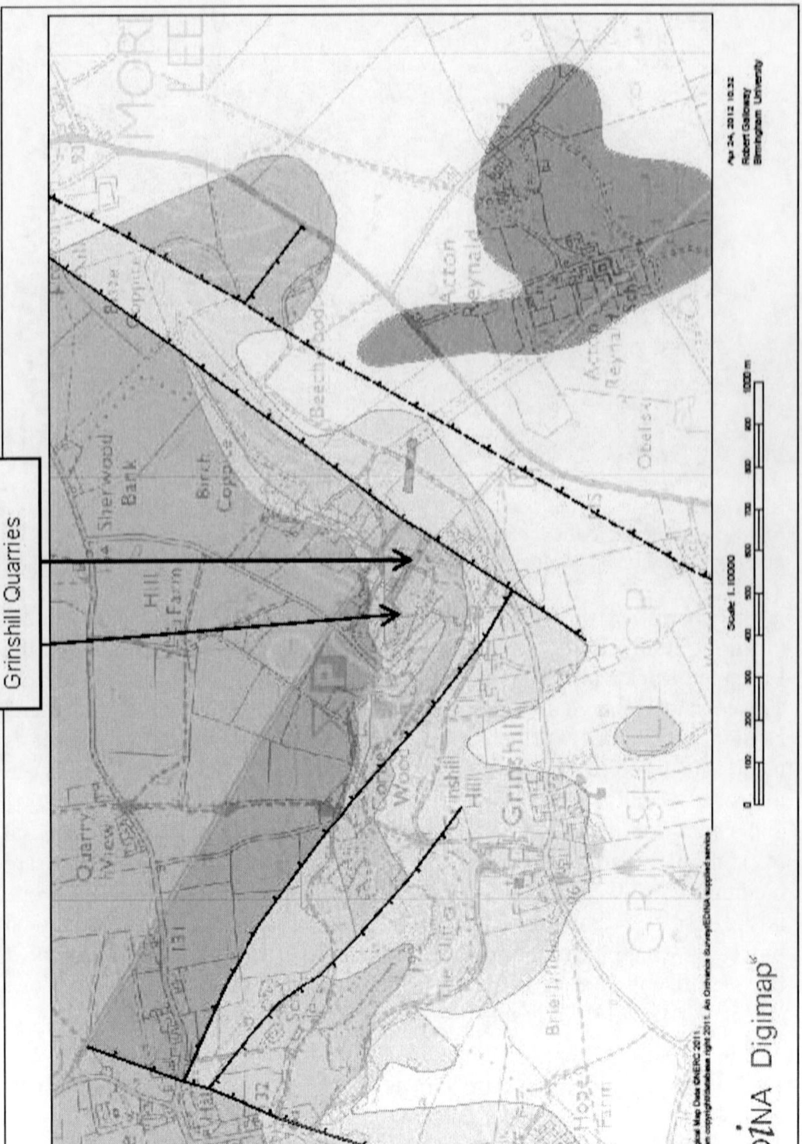

Figure 3.1: Geological map of Grinshill Hill. (Source: Geology Roam Digimap.edina.ac.uk/geologyroam, accessed 24/04/2012).

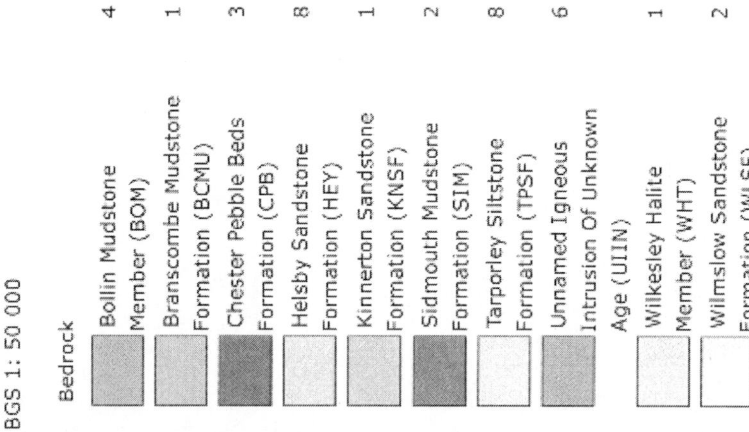

Figure 3.2: Geological Legend. (Source: Geology Roam. Digimap.edina.ac.uk/geologyroam, accessed 24/04/2012).

is the nearest of three faults, the other two are The Axial and Hodnet (Toghill, 2006). Below is a geological map and chemical analysis of the Freestone Rock shows that the mineral composition is highly siliceous and consequently chemically very resistant to weathering by atmospheric and erosion conditions. Further analysis of the sandstone shows that it is mostly made up of quartz in eight out of ten cases. Feldspar is present in the rock, which is a part of the rock-forming minerals (Bridge, *et al,* 2001 and Thompson, *et al,* 2008). The sandstone grains range from coarse 1.00 to 0.5mm to extremely fine0.125 to 0.063mm; however most grains are medium 0.5 to 0.25mm and fine 0.25 to 0.125mm in size. This sandstone is the result of 240 million years of physical, chemical and geological processes which have contributed to the different colours and have played the significant part in the density and strength of Grinshill Sandstone (Bridges, *et al,* 2001, Thompson, *et al,* 2008 and Watson, 1911).

3.3 History of Sandstone Workings at Grinshill Quarries

The dissolution of the monasteries between the years 1536 and 1541 initiated an enormous redistribution of lands. Tithes records belonging to the Parish of Grinshill show that the tithes formerly granted to the Abbots of Lilleshall Abbey, which also included the deeds to the Grinshill Quarries, were bought by local landowners. Sir Humphrey Lea purchased these tithes (tythes) in the Grinshill Parish (Greensell) after the dissolution of the monasteries. The Kilvert, Embrey and Wood families were other major landowners in Grinshill and have been for many centuries following the dissolution (Thompson, 2006 and Gough, 1981).

It was around the 14th century when quarrying commenced at Grinshill. However, the tower of Shrewsbury's first English Bridge, then known as the 'Old East Bridge', collapsed after a flood in 1546; the bridge was subsequently rebuilt with Grinshill stone. The Welsh Bridge, previously known as 'St. George's Bridge', was first built in 1155. Evidence of the quarrying of Grinshill Stone, between 1320 and 1330, is shown in its use in the construction of the refectory pulpit, once an integral part of the monks' accommodation in the former monastery buildings of Shrewsbury Abbey (Hannaford, 1996). Documentary evidence relating to possible secular quarry owners immediately following the dissolution is very limited. Major landowners possibly involved in the quarrying industry came from these families: Smith, Kilvert, Embury, Wright, Wood and Cureton. There is no confirming evidence that the first Welsh Bridge was constructed from Grinshill stone; however in 1795 this stone was used to construct the second Welsh Bridge. An archaeological excavation identified a surviving abutment of the original medieval Welsh Bridge which lies some 45 metres from the present waterfront (Watson, *et al,* 2015). During a period of a thousand years there were three English Bridges constructed at this same site. The first was built in Norman times, concurrent with the founding of Shrewsbury Abbey. The second English Bridge, designed by John Gwynn of Shrewsbury in 1767, was started on the 9th June 1769 and completed in 1774 at the approximate cost of £16,000. In 1921 the present English Bridge, designed by W. Ward, the Borough Surveyor, was also constructed from Grinshill Stone (Baker, 2010, Thompson, 2004 and Blackwell, 1985).

A document dated 1722 gives the first written evidence of Grinshill stone being used in nothing work carried out by a Mr Ric (Richard) Brasy for a Mr Jon (Jonathon) Dickins for 9 tuns (tons) and 10 hundredweight of stone from Grinshill for a new portico (Italian for porch leading to the entrance of a building); the stone cost 4s-6d, [total] ling £2-2s-9d (Shropshire Archives, Dickens to Hill, document no. 112/1/2380, accessed 12/10/2011).

In 1781, an Act of Parliament was passed for the Dividing and Inclosing of Common or Waste Lands in the Townships of Grinshill, Sansaw and Clive in the Parish of Grinshill and Saint Mary in the County of Salop. The Act names Richard Hill, Esq as Lord of the Manor of Hardwick, and states that within the manor there are several common or waste lands in the townships of Grinshill and Sansaw. It points out that Richard Hill, has, or claims, or is entitled to soils of all, and the singular, the commons or waste lands in the said Township of Grinshill (Shropshire Archives, Dividing and Inclosing Act, document no. 322/9/2, accessed 12/10/2011). The document continues to name John Garder, Robert Embrey and Sir Rowland Hill, Baronet and Dame Mary his wife; Andrew Corbet, Esp., Henry Cecil, Esq, and Emma his wife, John Wood, Gentleman and several other persons. These people were owners and proprietors of messuages, farms, lands and hereditaments within the same townships and their rights to common or waste lands (Shropshire Archives, Dividing and Inclosing Act, document no. 322/9/2, accessed 12/10/2011).

3.3.1 Carline and Linnel's Order Book

Carline and Linnel's Order Book for the 18th century, now stored in The Shropshire Archives, provides an insight into both the Grinshill Quarry operations and the purchasers of the quarried sandstone for building works. This is a fortunate and unique record, surviving from as early as 1771 to 1772, of the transactions undertaken by the quarry company and other individuals. It records dates from May 1771 to July 1807. The table is split into four boxes with a record of the names of the companies who purchased the sandstone, as well as of the transportation of the sandstone to individual locations. Other parts of the table show the cost of the sandstone per ton, the last section calculates the overall cost. From this table one can see that on 29th May 1771, 328 tonnes of quarried stone was taken from Grinshill Quarries at a cost of 25/- shilling per ton. The other four columns record the amounts of sandstone purchased and the cost per ton. The order book records people such as John Davis who on 20th July 1771 transported 62 tonnes of the quarried stone for Grinshill Quarries to a location not recorded. The cost of the transport was 9/-shilling per tonne and total cost. The tables also record, on the 31st August 1771, a Richard Evans transporting three loads of sandstone from Grinshill Quarries. It is interesting to note in the same column that the money was sent to his widow. The February 28th 1772 order book records that a canal barge and labourers were used to deliver 38 tonnes of quarried sandstone at a cost of 9/- shillings per tonne and in the final column is the total cost. 1772 is the first year in which barges on the canal are recorded as transporting sandstone from Grinshill Quarries. The last date of 20th July 1807 records stone required for a chimney piece. This order book of Carline and Linnel's is very valuable to the archaeological and historical records of

Shropshire. It shows the cost of the stone per tonne, the people who transported the quarried stone and what labourers were paid for their work. The order book uses the word 'credit' quite often, this could be for the credit of stone from the quarry, which was then paid at the end of the month (Shropshire Archives, Carline and Linnel's Order Book, document no. 800 box 20c, accessed 12/10/2011). Carline and Linnel's Order Books were very difficult to read, some of the writing was illegible, so in some places dots replace the missing words; the following table shows some of the orders:

Table 3.1: Carline and Linnel's Order Book, 1771 to 1772: material for the construction of Atcham Bridge:

DATES	DATA	SHILLINGS PER TON	£ : s: d:
29th May, 1771	328 tons of Grinshill and Build was Stone	@ 25/-	410: 1: 3
29th May, 1771	68 tons of Grinshill and Build was Stone	@ 16/8	55: 10: 0
29th May, 1771	263 tons of Grinshill and Build was Stone	@ 18/-	236: 18: 6
29th May, 1771	120 tons of Grinshill Stone…..at the New Inn	@11/-	66: 0: 0
29th May, 1771	267 tons of Grinshill Stone Bring and expense	@ 2/ 6	33: 8: 0
11th June, 1771	Mr. John Lowden, Dr Paid you at Grinshill Quarry on acc of getting stone ……	@	10: 10: 0
6th July, 1771	John Davis, Dr. Paid on Acc of 47 tons of Grinshill Stone from the quarry paid £ 21: 7: 5d	@ 9/-	Du 10: 10: 0
6th July, 1771	Richard Evens, Credit. 12.1 tons Carriage of stone from Grinshill Quarry.		5: 8: 5: ½
11th July, 1771	Mr. Lowden, Dr. For work on acc of work done at the Quarry paid you at Atcham.		5: 5: 0
20th July, 1771	Mr. John Machin. Dr. To each on acc of work done at the foundation of the 2nd Pier. From the …… acc		4: 4: 0 2: 2: 0
20th July, 1771	W.M. Haywood, Credit. To 3 days at Atcham and 3 days making drawings at Shrewsbury for Atcham.		1: 10: 0
3rd August, 1771	John Davis, Dr. To each on acc of stone carriage from Grinshill.		8: 8: 0
3rd August, 1771	John Davis, Credit. To carriage of 62 tons of stone from Grinshill Quarry.	@ 9/-	27: 18: 0

3rd August, 1771	Richard Evens, Credit. To carriage of stone 35 tons of stone from Grinshill Quarry.	@ 9/-	15: 17: 3
3rd and 6th August, 1771	Mr Lowden, Dr To cash on acc of work done at Grinshill Quarry. To [Ditto] cash self to the quarry by Davis.		5: 5: 0 10: 10: 0
10th August, 1771	Mr Richard Evens, Credit. To stone carriage from Grinshill 12: 2: 3 tons	@ 9/-	15: 9: 3
10th August, 1771	Mr. John Davis, Credit. To carriage of stone from Grinshill 14: 0: 0 tons.	@ 9/-	6: 6: 4
10th August, 1771	Francis Lee, Credit. 4: 7: 10 tons of stone carriage from Grinshill.	@ 9/-	1: 19: 5
10th August, 1771	Richard Morris, Credit. To 13: 18: 11 tons of stone carriage from Grinshill. To cash [Ditto] in full		6: 5: 6
10th August, 1771	Mr. John Davis, Credit To cash for stone carry from Grinshill in full		30: 7: 0
17th August, 1771	Richard Morris, Credit To 10: 15: 10 of stone carriage from Grinshill	@ 9/-	4: 17: 2 4: 17: 2
17th August, 1771	Mr. Richard Evens, Credit To carriage of stone from Grinshill 8: 12: 8 tons	@ 9/-	3: 17 8
17th August, 1771	Mr. John Davis, Credit To carriage of stone from Grinshill 10: 7: 7 tons		4: 7: 11
24th August, 1771	Richard Morris, Credit By carriage of stone 11: 8: 3 tons	@ 9/-	5: 2: 8
24th August, 1771	John Davis, Credit To loads of stone carriage from Grinshill 5: 4: 2 tons	@ 9/-	2: 11: 5

24th August, 1771	Mr. Haywood, Credit Six days self and horse		1: 10: 0 1: 15: 0
31st August, 1771	John Davis, Credit To three loads of stone carriage from Grinshill. By cash paid your wife on acc £ 4: 4: 0, £ 1:1:0 and £ 1:1:0	@ 9/-	6: 6: 0
31st August and 7th September, 1771	Richard Evens To three loads of stone carriage from Grinshill 6: 5: 0 tons. By cash sent …… your widow £4: 4: 0 = By cash sent your wife £ 2: 2: 0 = By cash sent your wife £ 1: 7: 0 =	@ 9/-	2: 16: 3 7: 13: 0
4th, 24th August and 7th, 14th and 21st September, 1771	Mr John Lowden, Credit By cash paid Masons, labourers as day acc, pumping out water and ……… the 3rd foundation to the 2nd Pier … removing ground from arch. Paid labourers at the ……………… To 14 days work of self	 @ 2/ 6 [Total]	2: 19: 8½ 5: 7: 8½ 18: 15: 11½ 1: 2: 5 1: 19: 6 1: 15: 0 32: 0: 3 ½
12th October, 1771	Mr. Haywood, Credit, To self and horse one day		5: 0
26th October, 1771	Mr. Richard Evens, Credit To carriage of 2: 4: 9 tons of stone. To cash paid you by wife	@ 9/- @ 9/-	19: 2 1: 12: 0
9th November, 1771	John Davis, Credit By carriage of 9: 18: 8 tons of stone	@ 9/-	4: 9: 5
16th November, 1771	The Justices of the County, Credit By cash of the Treasurer Mr. John Flint		60: 0: 0

16th November, 1771	Mr. Haywood By five days' work self and horse		1: 5: 0
28th December, 1771	John Davis, Credit By carriage of 9: 12: 2 tons of stone. By cash on acc [Ditto]	@ 9/- @ 9/-	4: 6: 5 5: 5: 0
28th December, 1771	Mr. Haywood By two days' work self and horse		5: 0
2nd January, 1772	The Justices of the County , Credit By cash of the Treasurer Mr. Flint		19: 0: 0
18th January, 1772	John Davis, Credit By one load of stone 2: 8: 7 tons carriage	@ 9/-	1: 1: 0
25th January, 1772	Richard Morris, Credit By carriage of 3: 18: 7 tons of stone to Atcham	@ 9/-	1: 5: 0
18th and 25th January, 1772	Mr. Haywood To self and horse two days' work		10: 0
15th February, 1772	John Davis, Credit By carriage of one load of stone 2: 7: 2 tons. By cash on acc [Ditto]	@ 9/- @ 9/-	1: 1: 0 2: 2: 0
28th February, 1772	Carriage Of Grinshill Stone from New Inn Paid Labourers and Barge. By 34 tons of stone delivered	@ 10/- @ 3/-	4: 9: 2 5: 2: 0
29th February, 1772	Mr. Richard Evans, Credit By carriage of stone 5: 19: 10 tons, three loads. To cash on acc by Mr. Evans	@ 9/-	2: 13 11 2: 2: 0
21st March, 1772	Carriage of Grinshill Stone from the New Inn. To Labourer s and Barge. Credit by carriage of 36 tons	 @ 3/-	3: 6: 11 5: 8: 0

25th March, 1772	Getting Rubble Stone By Labourer one day		1: 3
11th April, 1772	The Justices of the County, Credit By cash of the Treasurer Mr. Flint		30: 0: 0
25th October …. Poss 1800	Order Mr. Philips Jones for Mr Philips 3ft 6" Grinshill Hearth		
20th July, 1807	Grinshill Chimney Piece ordered by Mr. Tomkinson		

(Sources: Shropshire Archive, Carline and Linnel's Order Book, Document No. 800 box, 20c, accessed 12/10/2011).

This order book provides an opportunity to see recorded evidence of activities at Grinshill Quarries in the late 18th and early 19th centuries. The twenty-eight years documented evidence shows what amount of sandstone was purchased and how it was being transported to different parts of Shropshire. It records individuals who employed labourers and what they were paid. Transporting quarried sandstone was done by horse and cart, until the coming of the canals. This change is recorded in the order book showing that in 1772 canal barges started to be used, with paid labourers employed to move the stone. This unique order book records the different amount of money that Grinshill Quarries was charging at the time for their quarried stone. One has to appreciate that the quarry workers' pay would be included in the price of the limestone purchase from Grinshill Quarries. It also indicates the cost of sandstone per tonne in the years 1771 and 1772. In 1771 the sandstone was at 25/- shilling per tonne; to 3/- shillings in March 1772. It shows that stone transported by canal on a barge was cheaper. This order book, from the late 18th century, shows Grinshill Quarries as a busy site, excavating and processing sandstone, as well as the economic value of the mineral. The order book also provides information about the types of workmen such as, Mason and Labourers, who worked on the construction of Atcham Bridge. Overall the order book provides an excellent account of the amount of stone from Grinshill Quarries that was used in the construction of this bridge (Shropshire Archives, Carline and Linnel's Order Book, document no. 800 box 20c, accessed 12/10/2011). Below is a picture of Atcham Bridge:

Figure 3. 3: Atcham Bridge. (Source: www.bing.com/images/search?q=Atcham+ Bridge+shrewsbury accessed 23/08/2016).

3.3.2 Samuel and George Cank's Account Book for 1870 to 1873

Through my research at The Shropshire Archives, I was able to find a Samuel and George Cank Order Book for 1870 to 1873. It is unique to have two order books, this and Carline and Linnel's of 1771-1807, which are one hundred years apart. The former provides an insight into what was happening at Grinshill Quarries in the 1870's, it also shows that Bridge Quarries were operating at the time. Also, at the same time as this order book, in the late 19th century gas-powered, dressing machines were being used at Grinshill Quarries. It is interesting to note recorded in the Cank's order book that they use the term "stone got and loaded" at 4/- shilling per tonne. This is cheaper than the order book of 1771 to 1772 where to get the stone and transport it was 9/-shillings, whereas in Cank's order book it was 4/- 'get and load the stone and transportation' 4 ¾ shillings, so a combined total of 8 ¾ shillings per tonne. In the order book it also talks of work carried out at Grinshill Quarries by W. Townsend for £41: 12: 06, for the time this is quite a large amount of money. One can note that, recorded in this book, there are many more locations where the sandstone is being transported too. Transactions for the dates of the 30th July and 31st August 1870, show that transporting stone to: Yorton was 1 ¾ shillings per tonne, Hadnet, 5/-shillings per tonne, Albrightton, 3½ shilling per tonne and finally Salop, 4½ shilling per tonne. There are a few different places where stone is transported to; however, the places that are mentioned above the price of the transportation of the quarried stone remained the same over the three years of the order book. A difference in operation can be noted between the 1870 and 1873 order books, which show that at Bridge Quarry cranes were now at work. Appendix 2 shows the foundation of the cranes at Bridge Quarry. These cranes can also be seen in figures 3.4, 3.5 and 3.6. This order book records more detailed money transactions, samples of sandstone being sent to London for analysis, a Blacksmith's bill and wall building. The height of production in the quarrying industry was in the 19th century; from Samuel and George Cank's order book it seems that the Grinshill Quarries were economically sound and trade in limestone good.

In examining the ledger some of the terminologies used were, I suggest, particular to the 19th century such as: "Ridding, Tilling and Bur". "Ridding" means to clear the over-burden off the top of the solid rocks, this was normally achieved with picks, shovels and wheelbarrows; "Tilling" means working the cranes. The word "Bur" means bur-walling; this is the name for random walling, which is built in the manner (Trevor

Kirkham, per comm, 08/10/2011 and Shropshire Archives, The Account Book of Samuel and George Cank's business 1870 to 1873, document no. 4756/1/22, accessed 12/10/2011). The account book can be found in the appendix 1.

When comparing both ledgers which are over a hundred years apart, there are some similarities, for example the cost of extraction 'getting' and transportation of the stone. Getting and transporting stone for Grinshill Quarries in 1771 to 1772 was 9/- shillings per ton. From 1870 to 1873 the getting and loading of stone from Grinshill Quarries was 4/- shillings per ton. The Account Book of Samuel and George Cank's ledger shows that the stone used for the construction of Atcham Bridge cost 9/- shillings per ton, a difference of 5/- shillings per ton with the prices in 1870 to 1873. The 1870 to 1873 ledger gave no written evidence as to the intended use of the stone from Grinshill Stone Quarries. It is not known when Samuel and George Cank were proprietors of the quarries, or when they relinquished the lease. In 1860s the land around Bridge Quarries was being farmed. Today the same farmland is the site of Mount Pleasant business's (The Account Book of Samuel and George Cank's business 1870 to 1873, document no. 4756/1/22, and Downes, 1980).

Samuel Cank is recorded on the 1841 census, aged 15, as apprentice stonemason, but in 1851 and 1861 he is not recorded. A reason for this omission is that he could have moved, because in 1852 his son was born at Market Drayton. In 1862 Samuel Cank and a Thomas Smith owned a quarry in the Parish of Clive. From 1871 to 1874 Samuel Cank is recorded as quarry owner and contractor in Cassey's Directory. The 1891 census for Clive, states that Samuel Cank had retired aged 65 years, Samuel Cank had three sons who were stonemasons, George, Stephen, and another son named George; the records do not indicate why he had two sons named George (Thomson, 2004 and Historical Directories. www.historical directories.org, accessed 12/10/2011).

3.3.3 George Hancock and Son, proprietor of Grinshill Quarries around 1898

Prior to 1898 The Lord of the Manor, Sir Walter Corbet, of Morton Corbet, had worked the Grinshill Quarries for some time. From the ledger of Samuel and George Cank it was not easy to judge whether the Cank family was one of the possible owners of the various quarry sites at Grinshill Quarries. A document from Shropshire Archives named 'The Quarry', dated September 1898 mentions that Sir Walter Corbet had recently leased Grinshill Quarries to Messrs George Hancock and Son of

Shrewsbury. Once the Hancocks took control of the site, they set out to develop the quarries with a view to making them more economical, thus able to supply a larger market. For many years Mr George Hancock was managing director of The Bath Stone Firm, assisted by his son Mr John Lewington Hancock (Shropshire Archives, Sir Andrew and Acton Reynald, document no. 322/2/428/1, accessed 12/10/2011). Three pictures of either Bridge or Cureton Quarries, in or around 1898 can be seen below:

Figure 3.4: The Grinshill Quarries from their heyday in Victorian times untill1923 (Thompson, 2004).

Figure 3. 5: The Grinshill Quarries from their heyday in Victorian times until 1923 (Thompson, 2004).

Figure 3.6: The Grinshill Quarries from their heyday in Victorian times until 1923 (Thompson, 2004).

Grinshill Quarries submitted stone to Mr Kirkaldy to ascertain the crushing strain. Samples of Grinshill Stone were also taken from a block supplied to the engineer of The Elan Valley project, who had samples independently tested; the results proved very positive. The results of the tests showed that due to its high percentage of silica Grinshill Stone was very durable; and had the capability to withstand the polluted atmospheric conditions of large industrial towns, as well as the effects of saline found at coastal towns. In consequence the stone was much in demand as a dimension building stone (Shropshire Archives, Sir Andrew Corbet and Acton Reynald, document no. 322/2/428/1, accessed 12/10/2011). There are many examples of architectural features showing the durability of Grinshill Stone; including the following two examples. Firstly, the extant fabric of Morton Corbet Castle, built in the 15th century, located a few miles east of the quarries (Trevor Kirkham, per comm, 08/10/2011). The second example is the Old Grammar School, Shrewsbury, erected in the middle of the 17th century (Shropshire Archives, Sir Andrew Corbet and Acton Reynald, document no. 322/2/428/1, accessed 12/10/2011).

Late in the 19th century, stone dressing machines manufactured by Coulter and Harpin of Batley were used in the machine sheds at Grinshill. The machinery was capable of dressing stone 3.66 metres (12 feet) by 1.52 metres (5 feet). The machine was unique for its time, because when the apparatus neared the end of the block of stone, it became reversible. This avoided splitting the end of the blocks of stone. The machinery was powered by a Dowson Gas device. The shed also contained a saw-machine made by Beaumont and Co of Honley, near Leeds, used for sawing the larger blocks of stone. After the stone was prepared for market, it would be transported by horse-drawn wagons along track and then road to the railway station (Shropshire Archives, Sir Andrew Corbet and Acton Reynald, document no. 322/2/428/1, accessed 12/10/2011). Grinshill Stone Quarries sheds and machinery can be seen below:

Figure 3.7: The above picture above shows men working in the sheds possibly dressing stone; other men are working with a wooden crane which lifts stone blocks. Source: The Grinshill Quarries from their heyday in Victorian times until 1923 (Thomson, 2004).

Figure 3.8: the Grinshill Quarries from their heyday in Victorian times until 1923. (Thompson, 2004).

In the early 1900s Mr George Hancock and Son were still owners of The Bridge Quarries or Cureton Stone Quarries, as it was sometimes known. At this time there were a series of fatal accidents at the quarries.

3.4 New ownership of The Bridge or Cureton Stone Quarries 1923 to early 1980s

In 1921 proposals were made to widen The English Bridge, Shrewsbury, by the borough surveyor, Mr Arthur Ward, but whether the plan was accepted is not clear (Shropshire Archives, Sir Andrew Corbet and Acton Reynald, document no. 322/2/428/1, accessed 12/10/2011). It is possible that, at the time, stone from Bridge Quarry, Grinshill would have been considered, as the stone used in its original construction came from Grinshill. The Bridge or Cureton Quarries, now under the management of Mr John Levington Hancock, were failing and is reported to have closed in 1922 (Shropshire Achieves, Kelly's Directory, 1922). In 1923, the Hancock's were still in control of the quarries. There is conflicting documentation, as to who purchased the leases of Grinshill Stone Quarries in 1923 from the

Hancock Family; there is a reference to a Mr William Edwards, Mr Richard Smith and a Mr Charles Thursby as purchasers of the leases.

Further research shows that Mr William Edwards, Mr Richards and Mr Charles Thursby had formed a new company, known as The Grinshill Stone Quarries Ltd in 1923 (Shropshire Archives, Sir Andrew Corbet and Acton Reynald, document no. 322/2/428/1, accessed 12/10/2011 and Thomson, 2004). The document below is of The Grinshill Stone Quarries Ltd, shareholders in 1923, this is the only record information of the company. The table shows three people as directors in 1923 and the amount of shares per person. It also records their occupation at the time, Messrs, William Edwards, builder, Richard Smith, masonry contractor and Charles Pelham Thursby, gentleman. All the shareholders lived in Shrewsbury, it interesting to note that two of the shareholders have occupations, which is appropriate associated with the quarrying industry. Mr Charles Pelham Thursby descends from an old Shropshire family of great wealth and was considered a gentleman (Shropshire Archives, Sir Andrew Corbet and Acton Reynald, document no. 322/2/428/1, accessed 12/10/2011 and Thomson, 2004).

Table 3.2: The Grinshill Stone Quarries Ltd, shareholders 1923:

Date	Name	Address	Occupation	Ordinary shares Nos.	Total share
14th July,1923	William Edwards.	Mardol Head, Shrewsbury	Builder	1-1000 & 3001-3050	1050
14th July, 1923	Richard Smith.	Ivydene, Clive, Shrewsbury	Masonry Contractor	1001-2000 & 3101-3151	1050
14th July, 1923	Charles Pelham Thursby.	Russel Place, Cross House, Shrewsbury.	Gentleman	2001-3000, 3051-3100 & 3651-4000	1400

(Source: Shropshire Archives, wages ledger for 1926-1929 and other documents no. 6057, accessed 12/10/2011 and Thompson, 2004).

A Bridge Committee was set up to oversee the work on The English Bridge and there seems to have been a consultation between The Bridge Committee and The Grinshill Stone Quarries Ltd, with suggestions being put forward from both sides. Once there had been acceptance of these recommendations in September 1924, a contract was eventually signed between The Grinshill Stone Quarries Ltd and The English Bridge Committee represented by Messrs. Muirhead, Macdonald Wilson & Co, Ltd of Westminster, on 6th April 1925. After this agreement, there seems to have been movement in the shares of The Grinshill Stone Quarries Ltd. On the 21st of August 1925, The English Bridge closed to traffic and dismantling of the bridge began, so that on the 1st of February 1926, the new foundation stone was laid. Shropshire Archives, contain illustrations of preparatory dismantling as well as reconstruction work to The English Bridge. On the 16th September 1927, work on The English Bridge, Shrewsbury was completed and reopened to the public (Shropshire Archives, wages ledger for 1926 to 1929, documents no. 6057, accessed 12/10/2011 and Thompson, 2004).

Table 3.3: The Grinshill Stone Quarries Ltd, shareholders 1924 to 1925

Date	Name	Address	Occupation	Share Numbers	Share From	Total Shares
29th June, 1925	Frank Hearn Shayler.	The Red House, The Mount, Shrewsbury.	Architect	3151-3650 & 501-950	New Stock, W. Edward.	850
... May, 1926.	Albert George Eccleston.	The Dingle, Whitchurch.	Solicitor	1001-1500	Richard Smith.	500
4th October, 1926.	Francis Milverton Drake.	Dana Chambers, Shrewsbury.	Architect	1501-1750	Richard Smith.	250
5th November, 1926.	Charles Stewart Asbury.	7 The Square, Shrewsbury.	Solicitor	41-80, 991-1000 & 3001-3030	William Edwards.	80
5th November, 1926.	Bernard Norman Watkiss	7 The Square, Shrewsbury.	Chartered Accountant	1-4 & 951-990	William Edwards.	80
23rd November, 1926.	James Renwick Riddell.	The Square, Shrewsbury.	Accountant.	81-100, 30313050 & 1551-1650	William Edwards, William Edwards and J. M. Drake.	140
.. March ,1927	Albert George Eccleston.	The Dingle, Whitchurch.	Solicitor	101-400	William Edwards.	300

(Source: Shropshire Archives, wages ledger for 1926-1929 and other documents no 6057 accessed 12/10/2011 and Thompson, 2004).

From further research into The Grinshill Stone Quarries Ltd, shows that in 1924 to 1925, only one year after the company was formed and new shares issued that other people were asked to buy shares in the company. As we know that a contract was signed between the representatives of the English Bridge Committee and The Grinshill Stone Quarries Ltd to for the work on the bridge. This is a good indicator why there was need for more shares in the company; these new shareholders would provide additional finance that the company required. If one examines the table in detail the people that are now purchasing shares in the company are not men from the building trade. They are men from a different social level, solicitor, architects and accountants, which suggest they are men of money. In addition because they are solicitor, architects and accountants, which all have a profession that could be beneficial to the company.

The Grinshill Stone Quarries Ltd, wages Ledger from April 1926 to October 1929 was discovered, showing the chief shareholder and company owner Mr Charles Pelham Thursby, the first time he is recorded as being the company owner. This unique ledger records the rates of pay from the Mr Charles Pelham Thursby, the Managing Director, the supervisor of the masons, the chief masons being paid 1s 10d per hour, the senior mason pay is 2d and 3d difference per hour. The rates of pay differ for the ordinary masons by 1d per hour. The Blacksmith rate per hour 1s 4d and if one notes that the hours he worked 61.5 hours more than any other profession recorded; the driver worked is 59 hours for 1s 2d per hour, the driver worked just 2.5 hours less than the Blacksmith. The next three occupations differ the skilled quarrymen are paid 1s 3d and 1s 2d per hour, unskilled quarrymen 1s 1d, labourers and steam crane driver get 1s 0d per hour. The last two categories are the apprentices' gets 9s and wages clerk gets 10s per week. The ledger records seniority the professions and the different of wages paid and the long hours they worked. It also gives the amount of people working in each category and their names. It is interesting to note in the dress and comments column what the people had on for example dress-plus fours, dress-bower hat and cloth hat. The dress code distinguished the different working classes, like a social hierarchy. This ledger gives an insight of the pay, working and the hierarchy at work, which continued outside of work (Shropshire Archives, wages ledger for 1926 to 1929, documents no. 6057, accessed 12/10/2011 and Thompson, 2004). As with other documents, some of the pages were unreadable. A table below showing some of the details of the ledger:

Table 3.4: Grinshill Stone Quarries Ltd wages Ledger for 1926 to 1929

Job Description	Rates of Pay 1926-27	Hours Worked: normal and maximum	Number of People	Dress and Comments	Names of People
Chief Shareholder & Owner of the Company.	Paid weekly on Account of dividend £ 2.		1	Dress- Plus fours… Often does office work at 1s 0d per hour	Charles. P. Thursby.
Managing Director & Supervisor of Masons.	Paid a flat rate wage of £ 6: 5: 0 d per week.	Not Recorded	1	Dress-Bower Hat and 3 weeks holidays.	Richard Smith.
Chief Mason.	1s. 10d per hour.	49.5 & 59.5.	1	Dress-Bower Hat.	Freddie Downes.
Senior Masons.	1s. 8d per hour & 1s 7d per hour.	49.5 & 59.5.	4	Dress-Bowler Hat. Often worked away on site and paid railway fares.	H. Jones, C. Sellas, G. Barker and W. Dobbinson.
Masons	1s. 6d per hour.	46.5 & 46.5.	1	Dress-Bowler Hat.	Not Recorded.
Blacksmith.	1s. 4d per hour.	54.5 & 61.5.	1	Dress-Cloth Cap.	Robert. Sherwood.
Skilled Quarrymen.	1s. 3d & 1s. 2d per hour	49.5 & 54.5.	5	Dress-Cloth Cap.	R. Owen, H. Price, F. Forgham, Joe Chettoe and Joe Walford.
Driver.	1s. 2d per hour.	49.5 & 59.	1	Dress-Cloth Cap.	George. Smith.

Unskilled Quarrymen.	1s. 1d per hour.	49.5 & 51.5.	5	Dress-Cloth Cap.	A. Ashbury, R. Dytor, H. Mainwaring, John. Chettoe and A. Higginson.
Labourers.	1s. 0d per hour.	49.5 & 57.	12	Dress-Cloth Cap.	R. Greenway, H. Mainwaring, R. Mainwaring, Tom Davis, A. Evens, J. Purcell, G. Morgans,W. Reeve, H. Chettoe, A. Clawley, G. Walford and J. France.
Steam Crane Driver.	1s. 0d per hour.	49.5 & 50.5.	1		A. Evens: A Labourer who deputised when H. Thomas was laid off permanently.
Apprentice Masons.	9s. per week rising to at least £ 2. 0s.0d per week over 2 years.	49.5 & ….	3	Tea boy, wheel barrowed blunt tools to blacksmith.	F. Foster, R. Tipton, J. France, S. Austin and R. Newnes.
Wages Clerk.	10s. 0d per week.	Hours not recorded.	1	Replaced by voluntary worker Mrs. Richard Smith, in hard times.	

(Source: Shropshire Archives, wages ledger for 1926-1929 and other documents no. 6057, accessed 12/10/2011 and Thompson, 2004).

It is interesting to see that sometimes the Chief and Senior Masons had to work 59.5 hours per week, some 13 hours more than the ordinary masons. The blacksmith was considered to be a highly skilled workman and would sharpen the tools blunted by the quarrymen. They would also be responsible for making parts for machinery on the quarry site (Valley of Stone, www.valleyofstone.org.uk, 23/08/2016). This indicated the importance of the Blacksmith as no other worker is recorded completing so many hours; only the Chief and Senior Masons and the Driver worked approximately comparable hours. The records show that the Senior Masons often worked away from the quarry and were paid railway fares to travel to the different places. The wages clerk position seemed to have been temporary relying on the amount of business that The Grinshill Stone Quarries Ltd had; falling work would see the clerk replaced by Mr Smith's wife (Shropshire Archives, wages ledger for 1926-1929 and other documents no. 6057, accessed 12/10/2011).

The table 3.5 records the same categories of people as in the table 3.4. The differences are in the labourers section of table 3.5 shows there is now seventeen employed an increase of five from table 3.4. However, this falls to seven men in 1927, four in 1928 and one in 1929, which could have possible been the lorry driver, because he often worked as a labourer. The ledger shows a list and category of the employees from 9th July 1926 to 5th July 1929. Throughout this period only one Chief Mason is recorded and over that time the ledger indicates his wages were reduced by 1d, to 1s.9d in 1927. Table 3.6 indicates that, from 1926 onwards it appears the category of Senior Mason was discontinued. In 1927 and 1929 the same table shows that there were only three masons employed, a decrease from 1926, compared to 1928 when six masons were employed. This could indicate an increase in business at The Grinshill Stone Quarries Ltd in 1928. The category of labourers in 1926 shows seventeen men being employed (Shropshire Archives, wages ledger for 1926-1929 and other documents no. 6057, accessed 12/10/2011). The table below show the list and categories of employees:

Table 3.5: The Grinshill Stone Quarries Ltd, wages ledger 1926-1929 list of employees:

Category.	Rates of pay.	9th July, 1926.	8th July, 1927.	5th July, 1928.	5th July, 1929.
Owner	Not recorded	1	1	1	1
Managing Director	£ 6: 5s: 0d per week.	1	1	1	1
Chief Mason	1s.10d per hour then reduced to 1s. 9d per hour on the 11th November, 1927.	1	1	1	1
Masons	1s. 8d to 1s. 6d per hour.	4	3	6	3
Apprentice Masons	Fixed wage to 9s. to £ 2: 0s.0d per week.	3	2	2	2
Blacksmith	1s.4d to 1s.5d per week.	1	1	1	1
Skilled Quarrymen	1s.2d to 1s.3d per week.	5	3	3	3
Lorry Driver	1s.2d	1	1	1	1
Unskilled Quarrymen	1s.1d	5	4	3	1
Labourers	1s.0d	17	7	4	1
Crane Driver	1s.0d	1	1	0	0
Office Clerk	10s.0d per week and for a short time £ 1.	1	1	1	1
Total Workforce.		41	26	24	16

(Source: Shropshire Archives, wages ledger for 1926-1929 and other documents no. 6057, accessed 12/10/2011 and Thompson, 2004).

At Shropshire Archives there are also invoices belonging to The Grinshill Stone Quarries Ltd for 1922 to 1938. These documents provide evidence of the state of the company at this time. In 1922 Mr John Levington Hancock was still in control of Grinshill Stone Quarries, but the company was clearly failing by then as evidenced by the poor quality of book-keeping compared to previous ledgers. Invoices for the period September 1922 to the end of July 1923 give an insight to the decline of the company. These invoices reveal business was very slow, goods worth only £1065 being traded. Most of that business was of a masonry nature for Messrs Melluish Brother's Ltd of Cardiff, who were carrying out work on St. Samson's Church £186:15:7d, Messrs Wilcock & Co Ltd of Wolverhampton £140:14:11d, Messrs Harding of Nantwich £66:11:7d, Messrs Morris & Co of Shrewsbury and Messrs J. Sheppard of Apsley Street, Redditch £78:3:8:½. It is interesting to see from the invoices that work only amounted to £50 for Shropshire County Council. However, from Messrs Gerrard & Sons they had work worth £31:4:3d, but had a large contract for £851:5:8d for work to The Memorial Infirmary, Wrexham. This contract was taken over by The Grinshill Stone Quarries Ltd (Shropshire Archives, invoices for 1923-1937 and other documents no. 69057, accessed 12/10/2011). A small portion of The Bridge and Cureton Stone Quarries business related to gardens walls and flaggings; most of this work was carried out locally in the Shrewsbury area. There were two main contractors, Messrs Jervis of Wem for £72:18:3d and Messrs Derrington & Sons of Birmingham for £54:10:3d; the rest was made up of small transactions (Shropshire Archives, invoices for 1923-1937 and other documents no. 6057/2/2 to 6057/2/13, accessed 12/10/2011).

Mr Charles Thursby was senior shareholder and remained firmly in charge of The Grinshill Stone Quarries Ltd until his death in 1978. Mr Charles Thursby's early years in control showed modest profits in 1950 to 1951, but the figures fell in 1952 and in 1953 the results for that year were so bad that they were not recorded at the annual meeting. At this time the company applied to The Sandstone Industry Compensation Fund for help and financial support to pay the wages of their employees (Thompson, *et al,* 1928 and Downes, 1980). Then from 1954 the company made progress at a modest pace. What is interesting to note is that the company needed to increase its insurance cover, because of public of liability Grinshill Quarries. The Grinshill Stone Quarries Ltd was now under the control of (Thompson, *et al, 2008).* November 1968 saw the election of another director a Mr Trevor Kirkham. In May 1969 and October 1977, Mr Charles Thursby transferred most of his shares to his son, Mr Jim Thursby to add his own shares, which gave him financial control of the company. The death of Mr Charles Thursby in 1978 left three active directors; Mr Jim Thursby, Mr Trevor Kirkham and Mr Charles Thursby's widow running the company. In 1980, Mr Jim Thursby sold 1000 shares in the company to a Mr John O'Hare. With Mr Charles Thursby now gone from the scene, disagreements and frictions between the directors started to emerge. Between 1980 and 1983 these disagreements were aired at meetings of the old company and their solicitors. Sometime between these years a decision was reached to lease Grinshill Quarries. The Grinshill Stone Quarries Ltd was now under the control of Mr Jim Thursby and Mr John O'Hare (Thompson, *et al, 2008* and Downes, 1980).

3.5 Mr Jim Thursby and Mr John O'Hare owners of Grinshill Stone Quarries 1982 to 1989

The Grinshill Stone Quarries Ltd was leased under the following terms: The Grinshill Stone Quarries Ltd (1923) were to receive a royalty or rent at £10 per ton, with a minimum sum of £2500 to be paid yearly. The lease for the quarries would run for 10 years (Thompson, *et al, 2008* and Downes, 1980). The company in the 1980s saw a resurgence in its business. This came about by the re-establishment of equipment to prepare stone on site, such as sawing machine, which would enable the new company to supply builders, stonemasons and other contractors that needed the stone cut to size. It was interesting to notice that the method used to extract large blocks of stone was different from other quarries that I had researched. The new owners of Grinshill Stone Quarries Ltd used the Bristar power method, which secured large blocks of stone (Kirkham, per comm, 08/10/2012). This method was an English invention marketed by the Japanese. The quarrymen drilled holes around the required blocks of stone, each hole is 35 centimetres apart and to any depth up to 4 metres. Bristar powder, an expanding gel, is then packed into the holes and primed with water. This causes a chemical reaction, which makes the material crystalize. This process generates a great deal of heat and incredible force of crystallisation, which is accompanied by considerable expansion. The process is usually done overnight. The result of the chemical reaction causes the stone to stress which causes internal fracturing along the lines of the least resistance such as, the drill holes (Degussal, www.procureit.com, accessed 28/03/2012). The use of the Bristar powder is environmentally friendly, it is a soundless and safe, non-explosive demolition agent. It does not cause any flying rock, noise, ground vibration, gas, dust or environmental pollution. With this new invention, heavy duty, modern machinery such as, caterpillars are then used to move the processed stone (Degussal, www.procureit.com, accessed 28/03/2012).

In 1989 Mr Jim Thursby and Mr John O'Hare sold The Grinshill Stone Quarries to English Clay Ltd. They owned the company for nine years until they sold it to Mr John O'Hare in 1998. Mr John O'Hare and his wife Mrs Kathryn are now to date both directors of Grinshill Stone Quarries Ltd (Thompson, *et al,* 2008 and Downes, 1980). After telephoning and emailing Grinshill Stone Quarries Ltd, I was unable to arrange a meeting between Mr John O'Hare and myself. Mr Trevor Kirkham showed me around Grinshill Stone Quarries and I asked him what the stone was used for today; because of its location the quarry is unable to expand in any direction. Trevor replied, that there is enough stone left for restoration work, but for a complete commercial development there might not be sufficient stone (Kirkham, per comm, 08/10/2012). The stone from Grinshill Stone Quarries is considered to be as much in demand in the 21st century as it was in the past and still valued by architects. Whether the quarry has the quantity of stone to undertake large projects is another question (Grinshill Sandstone, www.grinshillstonequarries.co.uk, accessed 29/03/2012 and Kirkham, per comm, 08/10/2012).

3.6 Corbett Wood tracks and their uses

In the early 20th century men working at The Grinshill Stone Quarries Ltd still walked to work along many converging tracks, as they had been doing for many centuries. Nevertheless, there were quarryman's cottages in Corbett Wood. This was because, with a 12-hour working day, at the end of the day it was sensible to stay in these cottages all week and go home at the weekends (Downes, 1980). In 1861, some 51 men were employed at the quarries; ten years later in 1871 only 45 men were employed. Of the 45 men, 30 resided in the Parish of Clive and 15 in the Parish of Grinshill. Before this time men would walk the 4.3 miles or 7 km from Broughton-with-Wem to The Grinshill Stone Quarries Ltd. Between 1861 to 1871 the workforce decreased, which may have been related to the introduction of steam powered cranes at the Grinshill quarry. During my site visit, which included the working quarry called Mount Pleasant at Grinshill Hill, I was able to see features of previous workings and foundations for cranes at the former Bridge or Cureton Stone Quarries and in Corbett Wood; photographs of this can be seen in Appendix 2. Waggon tracks running through Corbett Wood from Bridge or Cureton Quarries to the villages of Clive and Grinshill reflected the pattern and importance of the quarries. One of the tracks showed the groves made in the sandstone to help the horses grip the ground, to stop the horses slipping down the slope with a heavy load. Teams of horses would sometimes be placed at the back of the wagon-trains in order to slow down the momentum; making it safer for both men and horses; photographs can be seen in the Appendix 3 (Kirkham, per comm, 08/10/2011 and Downes, 1980).

3.7 Shropshire County Council and Grinshill Parish Council Planning Concerns

The Grinshill Stone Quarries Ltd in 2001 applied for modification of condition 17 (B) of planning permission MN92/0783/NS to allow for removal and replacement of a quarry bench (application no. MN/2001/0768/NS). In a site visit by The Grinshill Parish Council on the 6th October 2001 they strongly expressed their opposition to the proposal. They raised concern about public safety if the proposal went ahead. The removal of the bench on the south face would produce a near vertical drop, which was near to the public access to Corbet Wood. They were also concerned about the inadequate fencing on the south side of the quarry. This was still apparent today after my own site visit on the 8th October 2011 not only on the south side, but nearly all around the old and present quarries (Shropshire County Council, Planning Committee, 10th October MN/2001/0768/NS, accessed 12/10/2011 and Grinshill Parish Council, documents, accessed 12/10/2001).

3.7.1 Application to Shropshire County Council by Grinshill Stone Quarries Ltd, for Extension of Duration of Planning Consent No. MN2008/1387/NS

Recently Grinshill Stone Quarries Ltd applied to Shropshire County Council to extend the life of the quarry. The application was passed to the Planning Committee. This application again caused concerns to be raised by The Grinshill Parish Council and a public meeting was held on 29/08/2008. Grinshill Stone Quarries Ltd wanted a 25 year extension to the life of the quarries, but the Grinshill Parish Council considered that 15 years was more acceptable. This was taken into consideration by the Planning Committee and a 15 year extension was granted in November 2008 (Shropshire County Council, Planning Committee 4th November 2008, MN2008/1387/NS, accessed 12/11/2011 and Grinshill Parish Council, documents, accessed 12/10/2001).

A second point raised by the Parish Council was why Grinshill Stone Quarries Ltd were considering changing the way they used explosive to extract stone blocks from the quarries. They could not understand why they had changed from the Bristar Powder method to the Cardax tube; for Grinshill Stone Quarries Ltd one answer could be the cost (Shropshire County Council, Planning Committee 4th November 2008, MN2008/1387/NS, accessed 12/11/2011 and Grinshill Parish Council, documents, accessed 12/10/2001. Again in 2008, as in 2001, concern was raised by the Grinshill Parish Council about the road access from Grinshill Stone Quarries Ltd. There was to be some improvement to the access road from the quarries via Sherwood Bank; until work was carried out the limit on stone transport would be only 2,500 tons per year (Grinshill Parish Council, documents, accessed 12/10/2001). Other points of concern were that geological groups could not be guaranteed access, because of safety concerns and related cost of insurance. It was noted that the Grinshill Stone quarries are an SSSI and it was questioned how interested groups could gain access in future. Any plans for the purchase of the former quarryman's cottage must take the form of a public auction or sale by tender, to achieve its market value (Grinshill Parish Council, documents, accessed 12/10/2001).

3.7.2 Shropshire County Council, Planning Committee

Grinshill Stone Quarries Ltd original application was for a 25 year extension; however this brought concerns from people living locally who thought this was too long. The local people considered that a period of 10 to 15 years would be more appropriate and could not see how any longer period of time was justified. The council considered all the options and the outcome of the application was that planning permission was granted for a period of 15 years, subject to certain conditions. The primary condition, that the Council required to be addressed, was access to and from the quarry via Sherwood Bank onto the A49. This was required under Section 106 of the Town and Country Planning Act 1990.

- New direction-finding from Sherwood Bank to the A 49 route
- £12,500 was needed for improvements on the approach road to the quarry from the A 49 at Preston Brockhurst.
- A system should be put into place for the maintenance of the verge adjacent to the Grinshill Stone Quarries Ltd entrance, which at the moment is being used as parking for the quarry employees.
- The parking area across from the quarry entrance shall not be used to park lorries, which operate from the Grinshill Quarry.

Consideration of all points had to be analysed before permission would be granted, such as environmental conditions and restoration of the site after conclusion of quarrying at Grinshill Stone Quarries Ltd (Shropshire County Council, Planning Committee 4th November 2008, MN2008/1387/NS, accessed 12/11/2011).

3.8 Quarry evolution: 19th and 20th centuries and History

An examination of old maps was undertaken to compare how the area containing the Grinshill Quarry had changed from the 19th century to date. The 1890 map of Grinshill Hill indicates that there were twelve quarries around or near to the hill. To the east of Clive Village there were three quarries. On the Grinshill (Hardwicke) boundary, around the base of Grinshill Hill, there were a further six quarries. To the north of Grinshill Village there were a further three quarries on what is the present quarry site. North-east of Grinshill Village there was an Old Clay Pit and two Old Sand Pits (www.digimap.edina.ac.uk/ancientroam, accessed 03/04/2014 and Shropshire County Council, www.shropshire.gov.uk/corbet-wood-management-plan).

The 1900 map of the same area shows nine quarries now on or around the hill, a reduction of three. The Old Clay Pit and the Old Sand Pits are still indicated on the 1900 map. A 1920 map of the same area now only shows three quarries remaining north of the Grinshill Village, near the boundary of Corbett Wood, a reduction of nine from the first map. On the 1950s map of the same area there are still three quarries recorded located on or around the boundary of Grinshill Hill, on the Clive Village side of the hill. The 1970 map of the same area of Grinshill Hill shows no quarries recorded as such, but the contours on the map clearly show the outlines of the present quarry. A 2008 map of Grinshill Hill records one quarry outside the red boundary mark of Corbett Wood and a second quarry inside (Digimap, www.ac.uk/ancientroam, accessed 03/04/2012 and Shropshire County Council, www.shropshire.gov.uk/corbet-wood-management-plan, accessed 04/08/2011).

Figure 3.9: Grinshill Quarries 1890. (Source: Digimap, www.ac.uk/ancientroam, accessed 03/04/2012).

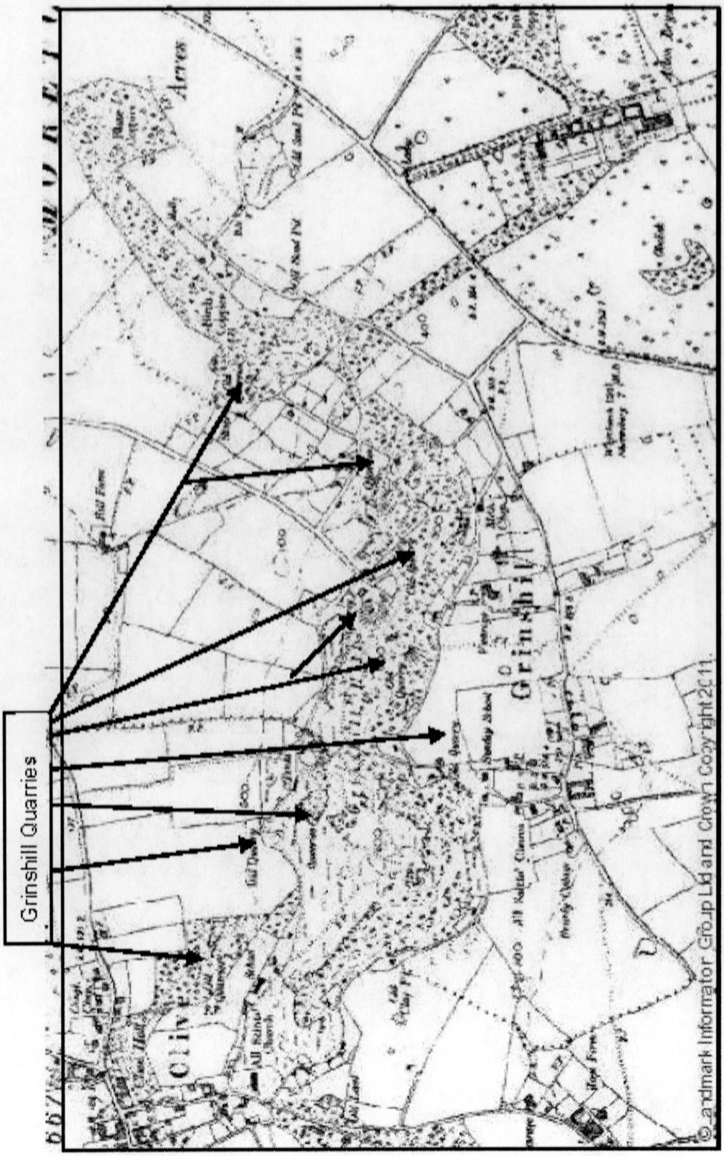

Figure 3.10: Grinshill Quarries 1900. (Source: Digimap, www.ac.uk/ancientroam, accessed 03/04/2012).

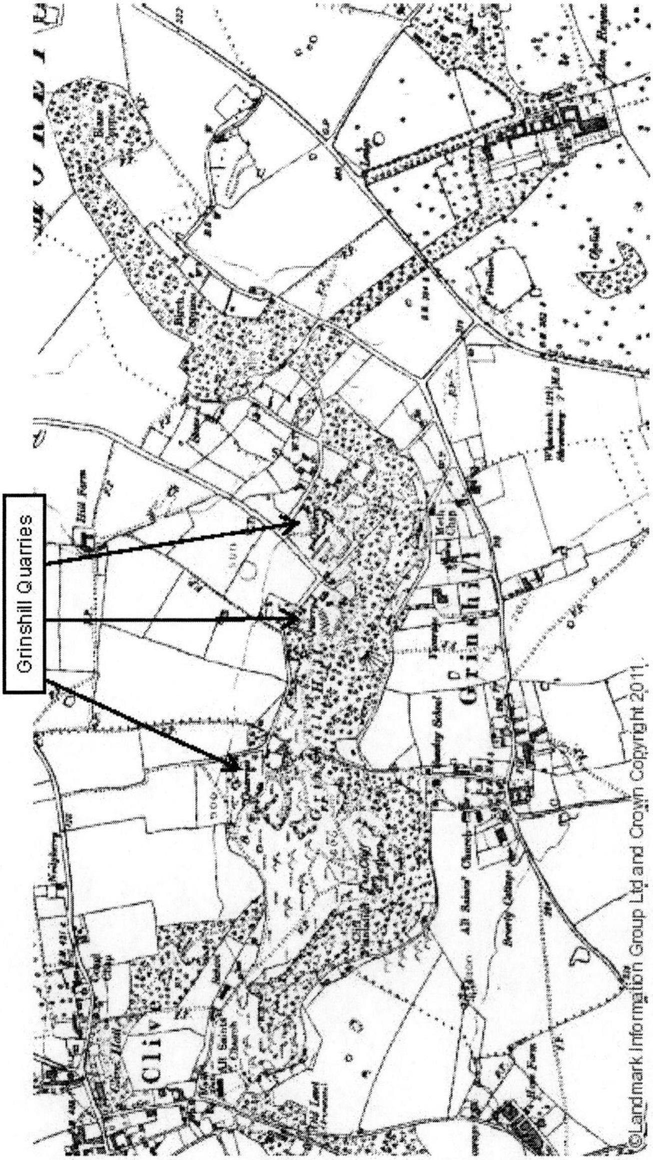

Figure 3.11: Grinshill Quarries 1920. (Source: Digimap, www.ac.uk/ancientroam, accessed 03/04/2012).

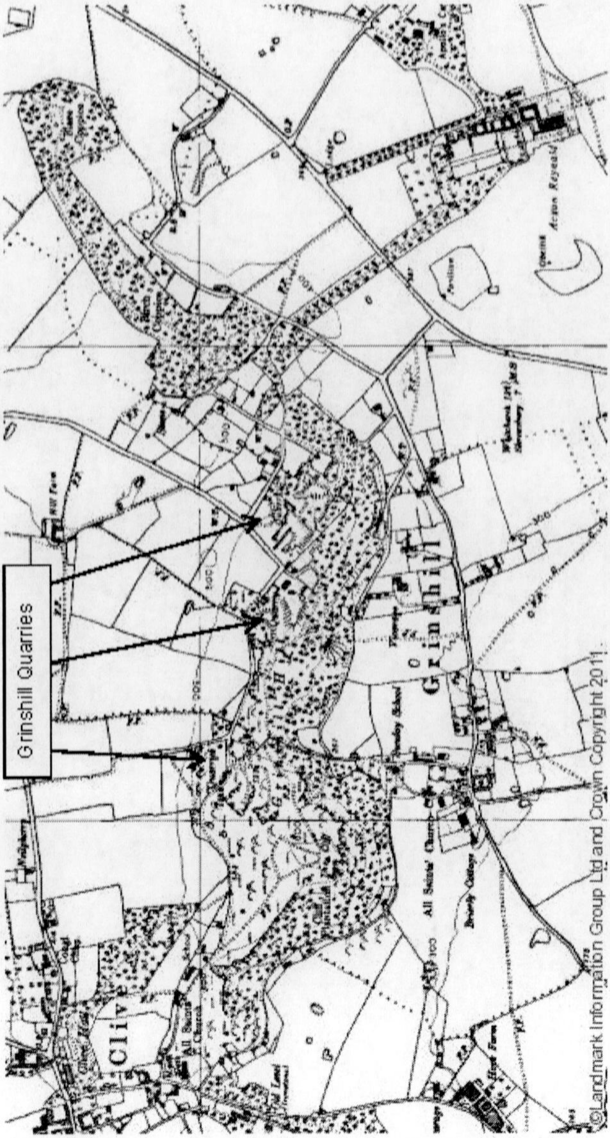

Figure 3.12: Grinshill Quarries 1950. (Source: Digimap, www.ac.uk/ancientroam, accessed 03/04/2012).

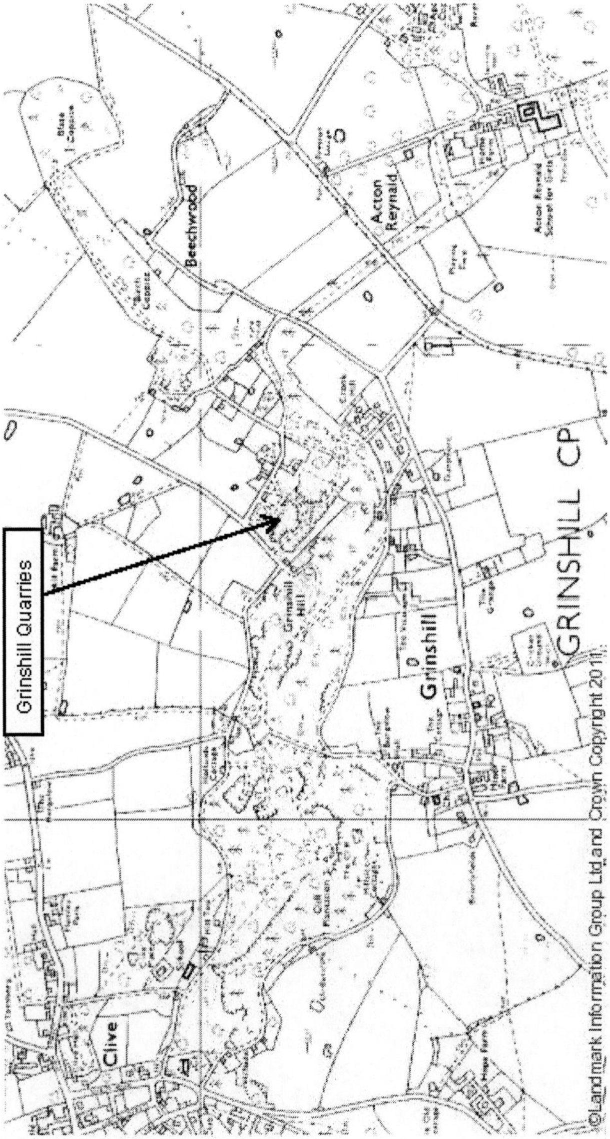

Figure 3.13: Grinshill Quarries 1970. (Source: Digimap, www.ac.uk/ancientroam, accessed 03/04/2012).

Figure 3.14: Corbett Wood: Source.
(Shropshire.gov.uk/Corbett-wood-management-plan, accessed 04/08/2011).

3.8.1 Planning History

There have been a number of planning applications made by Grinshill Stone Quarries Ltd over the years, the first being granted in December 1979 (N79/776) for the cutting of stone. This followed by a replacement office building permitted in September 1994 (MN194/0661/NS). The next and current planning permission was granted in December 1994 (MN92/0783/NS) for quarrying to take place for a further 15 years. Then an application was made by Grinshill Stone Quarries Ltd for an enhancement of the annual productivity from 4000 to 6000 tons per year, and granted in May 2000 (MN2000/0225/NS). This permission was on condition that a passing bay on the approach road to the quarry is built. This has not been completed and the permission has lapsed. Subsequently in October 2001 permission was granted for modification to an application (MN92/0783/NS). This was for the removal of a bench on the south-western side of the quarry (MN2001/0768/NS) (Shropshire County Council, Planning Committee 4th November 2008, MN2008/1387/NS, accessed 12/11/2011).

3.8.2 The Application

Application MN2008/1387/NS pointed out that quarrying had been carried out at Grinshill Quarry (Quarries) for many centuries. The stone quarried at this site produces the highest quality of stone in the county and has been used in the construction of buildings and in the restoration of historical monuments across

Shropshire and the Country (Shropshire County Council, Planning Committee 4th November 2008, MN2008/1387/NS, accessed 12/11/2011). The company states that the current 2.6 hectare, which is used for quarrying at present, is very small. In addition, there are working difficulties because of the restricted size of the quarry. The consequences of large tons of stone waste placed around the quarry edges and around the quarry floor over many years has caused hazardous working conditions. This overburden presents many problems, which affects efficient working and makes the quarry very unsightly. The removal of the waste stone would make working conditions safer and more efficient, as well as making the quarry more attractive to the eye for walkers (Shropshire County Council, Planning Committee 4th November 2008, MN2008/ 1387/NS, accessed 12/11/2011).

3.8.3 Site of Special Scientific Interest

Grinshill Stone Quarries Ltd is a site of geological SSSI and access would be appreciated for educational and research purposes. The 2.6 hectare that Grinshill Quarry occupies is located approximately 600 metres north-east of Grinshill Village. A plan of the complete area of Grinshill Hill, shows SSSI and Wildlife areas on and around Grinshill Hill, which incorporate Grinshill Quarry a SSSI site (Shropshire County Council, Planning Committee 4th November 2008, MN2008/1387/NS, accessed 12/11/2011 and Shropshire County Council, www.shropshire.gov.uk/ corbet-wood-management-plan, accessed 04/08/2011).

When restoration takes place, Grinshill Quarry's special status of an SSSI will be taken into consideration, particularly because of the discovery of rare fossil Rhynchosaurus bones and footprints of a Triassic reptile. Discovered on the same site were a second set of footprints, which are believed to belong to a reptile known as Cheirotherium (Toghill, 2006, (Shropshire County Council, Planning Committee 4th November 2008, MN2008/1387/NS, accessed 12/11/2011 and Shropshire County Council, www.shropshire.gov.uk/corbet-wood-management-plan, accessed 4/08/2011).

3.8.4 Mineral Polices

The Minerals Policy Statement 2 recognises the importance of this stone for building restoration and conservation of historic buildings (Local Government, Mineral Policy Statement 1 and 2 www.communitees.gov.uk, accessed 19/04/2012). Their policies encourage the reuse of building stone where realistically possible when restoring historic buildings. It also acknowledges the need for the extraction of stone from the Grinshill Quarry on a small scale for that use (Shropshire County Council, Planning Committee 4th November 2008, MN2008/1387/NS, accessed 12/11/2011). The company should guarantee sustainable use of the resources and circumvent any quarry sterilisation of stone caused by stockpiling of waste. To make the site an economically viable operation the proprietors are encouraged to identify, where feasible, material which can be set aside until it is required (Shropshire County Council, Planning Committee 4th November 2008, MN2008/1387/NS, accessed 12/11/2011 and Mineral Extraction and the Historic Environment. www.helm.org.uk. accessed 02/12/2011).

3.8.5 Consultations feedback

Different interested bodies held consultations concerning the effects of quarrying undertaken by Grinshill Stone Quarries Ltd regarding the possible extension of the quarry. Grinshill Parish Council raised many concerns but after voicing and having them addressed in the application, they formally expressed no objection to the extension of the life of the Grinshill Quarry (Shropshire County Council, Planning Committee 4th November 2008, MN2008/1387/NS, accessed 12/11/2011 and Department for Communities and Local Government, Mineral Policy Statement 1 and 2 www.communitees.gov.uk, accessed 19/04/2012).

3.8.6 Campaign to Protect Rural England, Shropshire Branch

The Campaign to Protect Rural England, Shropshire Branch (CPRE) was consulted and after consideration made a formal objection to application of Grinshill Stone Quarries Ltd. The reasons for their objections were that some dwellings are particularly close to the quarry workings and consideration should be taken regarding the noise from the workings; in particular when the noise levels reach 61 decibels. Concern was also raised about general operations, such as drilling and when the pecker is used. CPRE recognise that the quarrying at Grinshill has been undertaken for centuries and it is important in the construction of many buildings. They acknowledge that stone quarried from the site would be needed to maintain present historic buildings. It also accepts that people who purchased the houses near to the quarries must have been conscious of the presence of quarrying and of the consequences that come from quarrying operations. The last point made by the CPRE is that they would expect extractions of stone quarrying to be undertaken in a manner responsible and respectful of the lives of people living around or nearby to Grinshill Stone Quarries Ltd (Campaign to Protect Rural England, Shropshire Branch, www.cpreshropshire.org.uk, accessed 06/04/2012).

3.8.7 Environment Agency

The Environment Agency has no objections, but made some comments of how oil, fuels and chemicals were stored and suggested that these components should be stored in suitable compounds. Additionally, they instructed that there is no discharge of unclean or contaminated drainage from the site, which could go into either groundwater or any surface waters, whether direct or via soakaways. It is understood that the Grinshill

Stone Quarries are sited on a major aquifer of high vulnerability to surface contamination (Shropshire County Council, Planning Committee 4th November 2008, MN2008/1387/NS, accessed 12/11/2011 and Department for Communities and Local Government, Mineral Policy Statement 1 and 2 www.communitees.gov.uk, accessed 19/04/2012).

3.8.8 Highways Requirements

The Highways Department have no objection to the application being granted, subject to the following obligatory conditions.

- Drivers of vehicles entering and leaving Grinshill Stone Quarries Ltd should do so by way of the A49 at Preston Brockhurst.
- Grinshill Stone Quarries Ltd output should not exceed 8,000 tonnes per year. The quarry owners were instructed to construct the required passing bays and make improvements to the highways.
- They also required that twenty vehicles be the limit per day leaving and returning to Grinshill Stone Quarries Ltd, whether they were transporting excavated stone, or overburden stone from the site.
- They also requested that the management at the quarry introduce measures to record the type of vehicles leaving and entering the site, as well as a note being made of the tonnage of every vehicle. This should apply for the life of the planning permission and should be made available on request of the Mineral Planning Authority.

£12,500 was to be deposited with Shropshire County Council in order to assist with the highway improvements. This was required under Section 106 Agreement Order (Shropshire County Council, Planning Committee 4th November 2008, MN2008/1387/NS, accessed 12/11/2011).

3.8.9 Concerns of Private Properties nearby to Grinshill Stone Quarries Ltd

The nearest thirty-seven houses to Grinshill Stone Quarries Ltd were notified of the application for the quarry extension made on the part of the quarry company. Residents were concerned about the improvements to the access via Sherwood Bank being a way of justifying more intensive quarrying. There were questions about the restrictions on the number of lorries coming in and out of the quarry. They wanted to gain an understanding as to whether, or not, tractor loads of spoil would be covered by the same restrictions. Residents living near to the quarry suggested that Grinshill Stone Quarries Ltd had already been exceeding its current output limit for some time. However there seemed to be no evidence to back this concern raised by the residents. There were questions as to whether the access road at Sherwood Bank could cope with the increase of 8,000 tonnes per year (Shropshire County Council, Planning Committee 4th November 2008, MN2008/1387/NS, accessed 12/11/2011).

The increase in intensive workings at the quarry over the last few years is a factor that constantly comes to the surface during consultations with the residents. The conflict of interest occurs most obviously between the fact that local residents and visitors wish to enjoy the local amenities, such as Corbet Wood and its wildlife, as well as having the pleasure of their own back gardens and homes. They considered that all their concerns should be put above the need to extract quarried stone from Grinshill Stone Quarries Ltd. No reference has been made to the span of years that the quarry has been in existence at Grinshill and the part it has played in the construction of historic buildings locally and nationally over this continuous period (Blackwall, 1985 and Thompson, 2004, www.discover shropshire.org.uk , accessed 09/12/2011 and (Shropshire County Council, Planning Committee 4th November 2008, MN2008/1387/NS, accessed 12/11/2011).

3.8.10 Grinshill Stone Quarries Cottages Future Plans

Should The Grinshill Stone Quarries Ltd be allowed to sell the Quarryman's Cottage, then it must be with the consent of local and interested parties, who should also have the right to access it. People considered that the Quarryman's Cottage garden should cease to be used as a car park by the quarry company and must be included in any future plans for the cottage. It is suggested that both the cottage and the garden could be used for displaying the history of Grinshill and the flora and fauna of the area (Shropshire County Council, Planning Committee 4th November 2008, MN2008/1387/NS, accessed 12/11/2011).

Grinshill Stone Quarries Ltd have outlined in their application that they would seek to purchase from The Shropshire County Council the former Quarryman's Cottage situated south-west of the of the quarry workings. They would repair the building and use it as an office and a place to display their produce. Also part of the building would be made available, free of charge to any approved interested groups such as geological societies. A wall around the property would be built from Grinshill Stone with a gateway. Part of the cottage garden would be turned into a parking area for the company's employees; however elements of the original garden would remain (Department for Communities and Local Government, Mineral Policy Statement 1 www.communitees.gov.uk, accessed 19/04/2012 and Shropshire County Council, Planning Committee 4th November 2008, MN2008/1387/NS, accessed 12/11/2011).

There was strong opposition to the proposal for a secondary access to the quarry. It was thought that it would have an unfavourable effect on public areas and lessen the enjoyment of the local amenities such as the nature reserve and its ecology. The geological make up of Grinshill Hill and Corbet Wood recreational areas would be affected by HGV lorries driving through the area. It is suggested that these lorries would destroy the character of the historic woodland. The proposal of a secondary access from the A49 was considered to be too dangerous. There were also objections to the proposal to relocate the cutting shed from its present position to the quarry floor, because it would be closer to local residential housing (Shropshire County Council, Planning Committee 4th November 2008, MN2008/1387/NS, accessed 12/11/2011).

3.8.11 Appraisal of the Planned Expansion

The original planning permission was granted in December 1979 (N79/776) and was then reaffirmed by the current planning permission in December 1994 (MN92/0783/NS) which gave Grinshill Stone Quarries Ltd a further extension of 15 years life. It is believed that there are extensive stone reserves remaining and this is why the company have applied to extend the life of the operations. To the north of the quarry in close proximity are a number of residential properties. In the North Shropshire Local Plan the general area is protected against undesirable development. This is because of its recreational and touristic potential, as well as its landscape quality (Shropshire County Council, www. shropshire.gov.uk/corbetwoodmanagement-plan, accessed 04/08/2011 and Shropshire County Council, Planning Committee 4th November 2008, MN2008/1387/NS, accessed 12/11/2011).

3.8.12 The need for Grinshill Stone locally and nationally

The government guidance policy MPS1 recognises the importance of stone quarries and their role in the conservation and development of historic and prestigious buildings and bridges. The majority of other stones have distinct faults, some of which limit their ability to be cut and carved in particular directions. In the case of Grinshill Sandstone, it is homogeneous in structure and therefore of premium quality. This makes it capable of being cut in any direction, which produces regular shaped blocks of stone, carved into complicated shapes for example “freestone”. Grinshill Stone Quarries Ltd stated that there are permitted reserves in the order of 450,000 cubic metres (765,000 tonnes). They considered that seventy-five percent is capable of being used in restoration of historic buildings and new builds. (Department for Communities and Local Government, Mineral Policy Statement 1 and 2 www.communitees.gov.uk, accessed 19/04/2012, Shropshire County Council, Planning Committee 4th November 2008, MN2008/1387/NS, accessed 12/11/2011 and Mineral Extraction and the Historic Environment. www.helm.org.uk. accessed 02/12/2011).

3.9 Environmental Issues

Shropshire County Council recognized that there is a need to remove excess spoil heaps, but this must be weighed up against the limits put on the approach road to the quarry. Permission required that highway improvement works should be undertaken to accommodate the increase in output from 2,500 tonnes to 4,000 tonnes per year. The Planning Committee considered the company’s request and granted a three-year period; allowing the company to increase its output to 8,000 tonnes per year, then after this time limit they must revert to 4,000 tonnes per year (Department for Communities and Local Government, www.communitees.gov.uk, accessed 19/04/2012 and Shropshire County Council, Planning Committee 4th November 2008, MN2008/1387/NS, accessed 12/11/2011).

3.9.1 Noise

Local residents have raised objections about quarrying operations which include, in particular, the use of a stone mechanical breaker. Local residents alleged that the use of the mechanical breaker had increased over the past two years, when the quarry acquired its own breaker. The mechanical breaker is used to excavate narrow channels, perpendicular to the rock face, for the use of expanding powder (Bristar Powder), which is used to extract blocks of stone. The company required the permission of the Mineral Planning Authority before the mechanical breaker could be operated (Department for Communities and Local Government, Mineral Policy Statement 1 and 2 www.communitees.gov.uk, accessed 19/04/2012 and Shropshire County Council, Planning Committee 4th November 2008, MN2008/1387/NS, accessed 12/11/2011).

The Government's, Mineral Policy Statements 1 and 2, Local Government, Mineral Policy Statement 2 www.communitees.gov.uk, accessed 19/04/2012, provide advice on noise from mineral workings. The Mineral Policy Statement 2 instructs that areas with typical background noise characteristics of 45 decibels with an upper limit of $55dB_{LA90}$, is appropriate for normal quarrying operations. In more rural areas the noise levels from quarrying operations should not exceed more than 10 decibels. In reference to Grinshill Quarry operations, which have been long standing, it is considered that $55dB_{LA90}$ is an appropriate level for the control of noise at the quarry (Department for Communities and Local Government, Mineral Policy Statement 2 www.communitees.gov.uk, accessed 19/04/2012 and Shropshire County Council, Planning Committee 4th November 2008, MN2008/1387/NS, accessed 12/11/2011).

In 2007 a noise survey was carried out at the nearest property to Grinshill Stone Quarries called Harefoot Cottage. The survey indicated that the quarry was operating within the $55dB_{LAeq}$ noise levels laid down by the Mineral Planning Authority, with the exception of the mechanical breaker, which recorded noise levels between 42.7 and 60.4 decibels at Haresfoot Cottage. Since the survey was carried out, Grinshill Stone Quarries Ltd has installed further noise reduction to the mechanical breaker.

They also upgraded the chisel attached to the breaker, which works faster in cutting the stone and therefore reduces the operating time of the breaker. Grinshill Stone Quarries Ltd then commissioned a further noise survey in May 2008. This survey recorded noise levels between 43 and 55 decibels attributable to the mechanical breaker. In the 1 ½ hours that the noise survey took place, noises were also recorded from RAF helicopters based at Shawbury that flew over Grinshill with up to 62 decibels. At the same time the survey recorded noise levels from vehicles driving down the lane next to the quarry. This highlighted the additional noise levels present, but which are not associated with quarry operations. Following the attachment of a noise reducing fitment to the mechanical breaker, the noise levels now comply with the Mineral Planning Authority and The Government Guidelines (Department for Communities and Local Government, Mineral Policy Statement 2 www.communitees.gov.uk, accessed 19/04/2012 and Shropshire County Council, Planning Committee 4th November 2008, MN2008/1387/NS, accessed 12/11/2011).

Planning permission already in place, required that the doors to the cutting shed should be kept closed during sawing operation, at all times, except when vehicles are loading or unloading. The quarry company pointed out that the closing of the doors is not possible because of forklift vehicles coming in and out. The Planning ommittee therefore required the company to install acoustically attenuating plastic sleeves to the cutting shed doors. This would allow the forklift vehicles to move freely in and out of the cutting shed (Department for Communities and Local Government, Mineral Policy Statement 2 www.communitees.gov.uk, accessed 19/04/2012 and Shropshire County Council, Planning Committee 4th November 2008, MN2008/1387/NS, accessed 12/11/2011).

3.9.2 Dust

The Mineral Policy Statement 2 provides guidance for the control of dust associated with mineral workings. The policy recognises that if dust from surface workings is not managed, or controlled properly, it can have an environmental effect on the local community's way of life. Therefore, Local Planning Authorities should consider the effects of dust on the local environment and communities, before granting planning permission. There are no records of complaints from local residents about dust from Grinshill Stone Quarries Ltd and no concerns were raised in the planning consultation process. The quarry is surrounded by woodland and this helps to control the dust by trapping the majority within the quarry confines (Department for Communities and Local Government, Mineral Policy Statement 2 www.communitees.gov.uk, accessed 19/04/2012 and Shropshire County Council, Planning Committee 4th November 2008, MN2008/1387/NS, accessed 12/11/2011).

3.9.3 Waste Heaps

One resident raised concerns about the practice of storing waste above surrounding ground levels on the south-eastern periphery of the quarry. The piles of waste stone are surrounded and screened mainly on all sides by woodland. However the waste heaps can be visible from a garden of a property nearby. The proprietors of the Grinshill Quarry state that the waste stored on the site is approximately 36,000 tonnes; this is considerably less than previous operators of the quarry (Communities and Local Government, Mineral Policy Statement 1 and 2 www.communitees.gov.uk, accessed 19/04/2012 and Shropshire County Council, Planning Committee 4th November 2008, MN2008/1387/NS, accessed 12/11/2011). The proprietors of the quarry have put in a system to reduce the piles of waste; the material is processed to separate the stone from the sand. A result of this process is that the reclaimed stone can be cut and put on the market, while the screened spoil is sold on to be used as an agricultural surface material. This process is hampered by the output limits, which in turn restrict the removal of the wastes. Therefore current proposals seek to increase the output of the quarry and consequently reduce the waste heaps. Any increase in output should be over a three-year period only, which would incorporate the lowering of the waste heap sites (Communities and Local Government, Mineral Policy Statement 1 www.communitees.gov.uk, accessed 19/04/2012 and Shropshire County Council, Planning Committee 4th November 2008, MN2008/1387/NS, accessed 12/11/2011).

3.9.4 Extraction and Explosive Methods

For many years Grinshill Stone Ltd has been using Bristar Powder in the extraction of stone blocks from the quarry. They also have permission to use black powder explosives for extracting stone; however this sort of blasting is rarely used. The effects of using this method can result in damaging the stone. At the same time, concerns had been raised by local residents that black powder could have harmful effects on the stability on their properties (Shropshire County Council, Planning Committee 4th November 2008, MN2008/1387/NS, accessed 12/11/2011). An analysis was made of both the Bristar Powder and Cardox extraction methods. Bristar Powder is a non-explosive demolition agent; it is environmentally friendly and is soundless. As well as being noiseless, its other benefits are that it does not cause any flying rock, ground vibration, gas, dust or environmental pollution. Bristar Powder is ideal for all types of demolition work, particularly where quarrying is taking place close to residential properties, for example at Grinshill Quarries. Where other explosives cause noise, damage, pollution and inconvenience to surrounding environment, plus local residents, Bristar is noiseless and can be used overnight. Bristar Powder is easy to mix with a drill and paddle, needs no licence or special expertise, and is economical as it saves on manpower, production price, while at the same time making less waste, which is a major problem at Grinshill Quarries (Bristar, www.degussa.cc.co.uk, accessed 10/04/2012).

Cardox Tubes Explosive, similar to Bristar Powder, is an expanding tube within a tube sheet; it has an insert and detonation device. The insert is fitted coaxially within the tube and then the detonation device encircles the insert and can produce, upon detonation, an explosive force. When two Cardox tubes are used it can produce a split and leave a stone block of 50 tonnes. It is fully approved by the Health and Safety Executive Mines Inspectorate. It is also classified as a nonexplosive. No special permits are needed for storage or transportation. There is a reduction in the amount of fine waste caused by this process. Cardox can be used in a quarry when people are at work as they only need to move away during its use. There is virtually no dust, it is cost effective and fewer drill holes are needed to extract a block of stone. It has an indefinite shelf-life and causes no shock waves, or vibration to the quarry site or area. Also there is delay system available and Cardox tubes are refillable and can be reused for a period of up to 15 years (Cardox International Limited, www.cardox.co.uk, accessed 10/04/2012).

There are some similarities between the Bristar powder and Cardox tube systems. They both reduce waste stone in the quarry which is a welcome benefit to what is already a problem at Grinshill quarries. They are both environmentally friendly and cause no noise or vibrations; however the Cardox tube system does leave a little amount of dust, which the Bristar powder does not. One of the major differences is that Bristar powder is used at night, which is much better for local residents. The Cardox tubes system is used any time day or night, which gives local residents no rest from the quarrying activities. Another difference is the amount of drilled holes that is needed with the Cardox tubes system compared to the Bristar powder. Overall there is not much to choose between Bristar powder and Cardox tubes system. Grinshill Stone Quarries Ltd has said they are willing to continue to use the Bristar powder. Shropshire County Council, Planning Committee have said that any future planning permission would require the continual use of the Bristar powder method (Shropshire County Council, Planning Committee 4th November 2008, MN2008/1387/NS, accessed 12/11/2011 and Quarries and Ancillary Activities, www.environ.iq/on/publications, accessed 17/04/ 2012). Appendix 4 showing pictures of stone extracted by Bristar Powder.

3.9.5 Health and Safety

Grinshill Parish Council has raised concerns about the fencing around the quarry, particularly on the south-eastern side. There are Public Footpaths in Corbet Wood, some of which are very near to the quarry perimeter. On the day of my site visit to Grinshill quarries and my walks around Corbet Wood, it was clear to see the possible danger that could be caused by the inadequate fencing around the quarry edges. There were no Public signs indicating the quarry edges or the steep drops. Additionally there is an abandoned quarry within Corbett Wood, which also has inadequate old fencing around it; this old quarry is at least 10 metres deep. If new planning permission is granted then up-to-date fencing needs to be put into place (Department for Communities and Local Government, Mineral

Policy Statement I and 2 www.communitees.gov.uk, accessed 19/04/2012 and Shropshire County Council, Planning Committee 4th November 2008, MN2008/1387/NS, accessed 12/11/2011).

3.9.6 Restoration of the quarry site

Grinshill Stone Quarries Ltd presented a restoration plan with the current application. Their suggestions include: retention of quarry faces, building an access ramp, formation of a heath land habitat in the base of the quarry. However, within the quarry at present there are substantial stone reserves of which only a little amount of stone would be recovered in the permitted timescale. Nevertheless if quarrying operations were to cease prior to all the reserves being recovered, the proposed restoration plans put forward by the company could be modified (Department for Communities and Local Government, Mineral Policy Statement 1 www.communitees.gov.uk, accessed 19/04/2012, Shropshire County Council, Planning Committee 4th November 2008, MN2008/1387/NS, accessed 12/11/2011 and Mineral Extraction and the Historic Environment. www.helm.org.uk, accessed 02/12/2011).

Grinshill quarry is a geological SSSI because of the rare fossil finds and it is important that any restoration plans safeguard these geological features. Also within the quarry boundaries are flora and fauna which are considered of special interest. Similar plants of special interest are found on Grinshill Hill, so there is a wide area to be taken into consideration in the restoration plans for the quarry site and surrounding area (Natural England, www.naturalengland.org.uk/regionwest_midland, accessed, 12/01/2012). The proposed restoration plans include the retention of the rock faces on the northern side of the quarry. This would include any associated features such as, volcanic dykes and historic pillar workings. The southern side of the quarry spoil would be placed up to 25 metres thick against the old quarry faces to form a sloping access ramp. The company have consulted Natural England and have agreed on a code of practice, thus safeguarding the geological interests whilst the work is being carried out (Natural England, www.naturalengland.org.uk/regionwest_midland, accessed, 12/01/2012, Extraction and the Historic Environment. www.helm.org.uk. accessed 02/12/2011, Department for Communities and Local Government, Mineral Policy Statement 1 www.communitees.gov.uk, accessed 19/04/2012.

3.9.7 Groundwater

Grinshill quarry is located on the perimeter of a groundwater protected sector, but the company states in the application that quarrying operations would not give any rise to a potential for pollution. They state no fuels would be stored near to the protected area and that no unclean drainage would occur. Measures would be in place to intercept run-off water during heavy rain. Quarrying operations are restricted to 91.4 metres above ordnance datum, so pumping would not be necessary (Extraction and the Historic Environment. www.helm.org.uk. accessed 02/12/2011, Shropshire County Council, Planning Committee 4th November 2008, MN2008/1387/NS, accessed 12/11/2011).

3.10 Discussion

The Grinshill Quarries have been particularly interesting to research because of the amount of documentation available in the Shropshire Archives and the articles from the Clive and Grinshill Conservation Group. It was very interesting to be given a personal tour of The Grinshill Stone Quarries Ltd by Mr Trevor Kirkham who imparted his vast knowledge of the history of the quarries and surrounding areas. Grinshill stone has played a major part in the construction of buildings and bridges in Shropshire including St Chad's Church, Shrewsbury Station, Lord Hill's Column, The English and Welsh Bridges, Shrewsbury Library, Morton Corbett Castle, Badger Hall, The Manor House Grinshill, Clive Hall, Attingham Park and many more buildings in and around Shrewsbury. In other parts of Britain Grinshill stone was supplied for use in Downing Street, Chequers, Hereford Cathedral, Worcester Cathedral, Powis Castle, Warwick Castle and the University of Aberystwyth. It has been used in the restoration of historic buildings both in Shrewsbury and across the country. The Grinshill stone has been used in new buildings such as Coventry Cathedral and Sussex University (Thompson, 2005 and Downes, 1980).

It was interesting to be able to access documentation, for example ledgers from the 18th and 19th centuries, to analyse the difference in wages and the amount of stone supplied from Grinshill Quarries. These documents show that Grinshill Stone Quarries were at the height of production in the 18th and 19th centuries; it declined somewhat in the early 20th century, but is still producing quarried stone today.

What does the future hold for Grinshill Quarries with the constant restraints in place to protect the environment and the need to satisfy local residents' anxiety? Grinshill Stone is still needed for restoration of historic buildings and monuments locally and nationally. The quarry is surrounded by Corbett Wood, a wildlife site and SSSI of geological importance. There are safety issues that need to be addressed by Grinshill Stone Quarries Ltd and Shropshire Council regarding the state of the fencing around the quarry sites to protect the public from falling over the edge into certain quarry sites in Corbett Wood.

3.11 Conclusion

Grinshill Stone Quarries Ltd location in the early years of quarrying operation was no problem, but as legislation began to be altered around 1945, any quarrying expansion application would have been fruitless; this was confirmed in an interview with Mr Trevor Kirkham. Corbett Wood, which is now a wildlife site and SSSI of geological importance, contains the evidence of Grinshill's original deep quarrying operations. Adjoining the wood, the present day Grinshill Stone Quarries only have enough stone left for future restoration work. In 2008 the local people living near to Grinshill quarries were concerned about the possible extent and time of ongoing quarrying operations and the probable affect in would have on their lives and properties. This being said, the stone from Grinshill Stone quarries is an important historical resource that will be needed for the future conservation of many historic buildings in the county. It is clear that there are structural issues with keeping the quarry working which need to be addressed, such as: road access, too much waste and methods of working. However, the one thing in their favour is the Bristar Powder method used to extract stone from the quarry, which is very quiet and environmentally friendly.

CHAPTER 4

AGRICULTURAL AND BUILDING LIME: LLANYMYNECH HILL QUARRIES

4.1 Introduction

This chapter will investigate the historical and archaeological background of the former limestone quarries at Llanymynech Hill, Shropshire. The limestone from Llanymynech Hill was used for agricultural purposes, as was some of the limestone from other quarries such as Wenlock Edge and Lilleshall. Llanymynech lies on the border between Shropshire in England and the two counties of Powys and Montgomeryshire in Wales. Llanymynech Hill lies approximately five miles southwest of Oswestry. The major landowners in the Llanymynech area have been the Earl of Chirk, the Earl of Powis who took over the Chirk Castles Estate lands and The Earl of Bradford (the Bridgeman family). The former Llanymynech quarries and buildings now lie within a Heritage Centre and Heritage Area and the map below shows the extent of the protected area (Shropshire Archives, Landscape and Development Evaluation, document no, q FL 71.1, accessed 22/05/2013, Sherlock, *et al,* 2001 and Adams, 1970). Below is a map of the Llanymynech Heritage area:

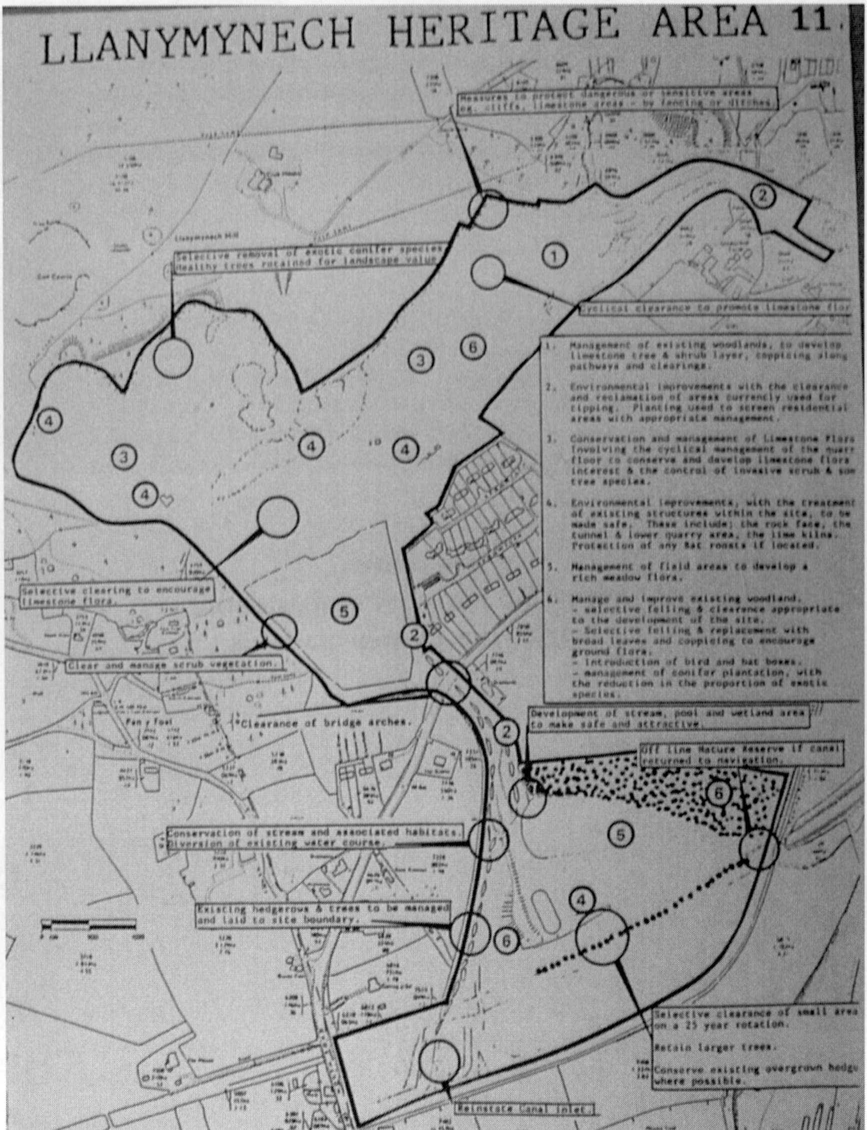

Figure 4.1: Llanymynech Heritage Area: (Sources Shropshire Archives Landscape and Development Evaluation, documents no, q FL 71.1, accessed 22/05/2013).

The word Llanymynech is completely Welsh in origin. It is considered to have derived from"Llan", the church or inclosure and "Mynech" a monk, which together means "The Church of the Monk". There is another meaning from a word "Mwynrw" which means the miner, "The Church of the Miner". Today the first meaning seems to be preferred, but the latter, I believe seems to have more significance, considering the amount of mining and quarrying that has taken place over the centuries (Joan Zorn, Per Comm, 04/04/2013 and Fewtrell, 1878, p 379).

4.2 Geology of Llanymynech Hill, Shropshire SJ264221

The prominent geological features of Llanymynech Hill and its district include the enormous deposits of carboniferous limestone. The carboniferous limestone of Llanymynech Hill was deposited approximately 290 to 360 million years ago. The limestone sequence on Llanymynech Hill is known as the Whitehaven Formation, or lower limestone. It is comprised of white rubbly limestone, which is well bedded and magnificently jointed (Fewtrell, 1878, Toghill, 2006, Rhodes, 2008). The limestone rock on Llanymynech Hill was created in a warm and shallow tropical sea, with a slender strip of land just to the south. This narrow strip of land was attached to an area known by geologists as St. George's Land which has now completely disappeared; eroded over time. Llanymynech Hill's limestone continues onto Prestatyn on the north Welsh coast. The limestone which forms the southern end of the west facing ridge runs northwards to Eglwyseg near Llangollen. The strata lies on folded Ordovician and Silurian rocks, which are divided into four different layers; working from the bottom:

1. Basal Shale is up to 30 metres in depth. This can be seen on the southeast side of Llanymynech Hill and are stained red to a depth of approximately 22 metres.
2. The Lower Limestone (Pant formation) is approximately 85 metres in depth. This is comprised of a white, jointed, rubbly limestone.
3. The Upper Grey Limestone (Nod Glas formation) is approximately 70 metres to 100 in metres in depth. This limestone has certain dark bituminous layers, comprised in part of large brachiopods.
4. The Sandy Limestone (Leete formation) is approximately 100 metres to 200 metres in depth (Fewtrell, 1878, Adams, 1970, Toghill, 2006 and Rhodes, 2008).

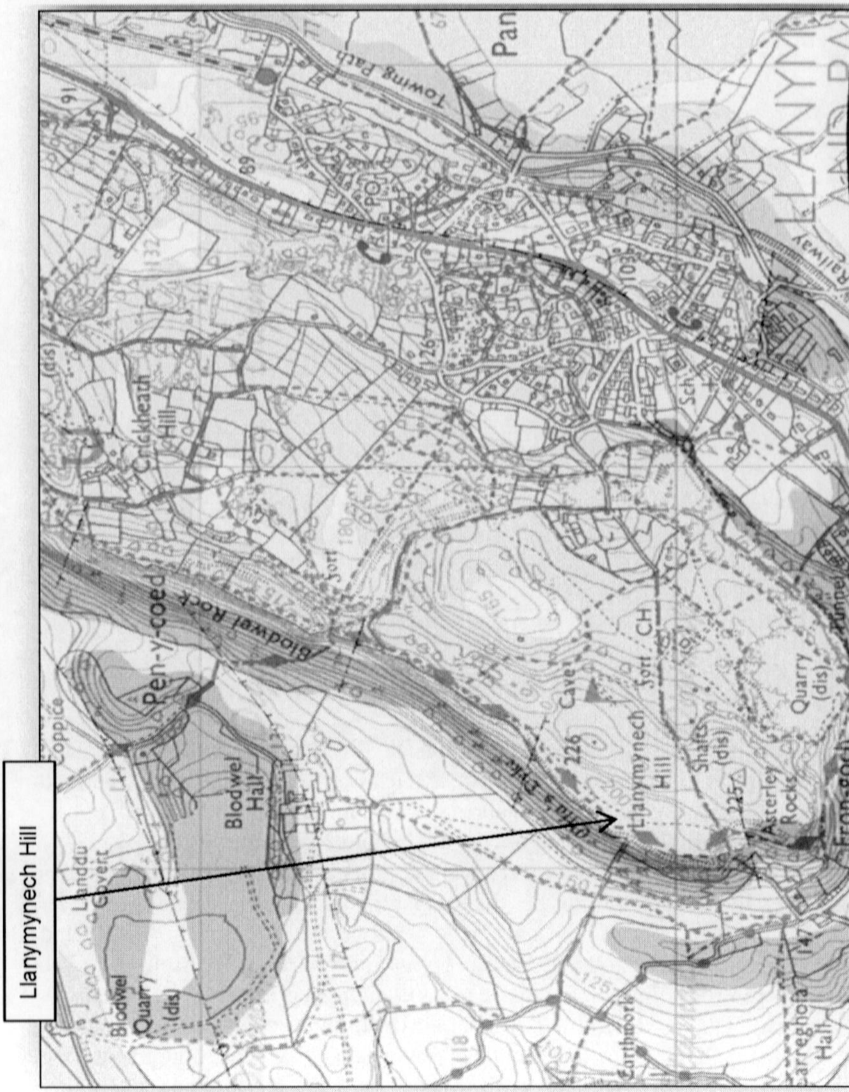

Figure 4.2: Geological map of Llanymynech Hill. (Source: Geology Roam. Digimap.edina.ac.uk/geologyroam, accessed 12/06/2013).

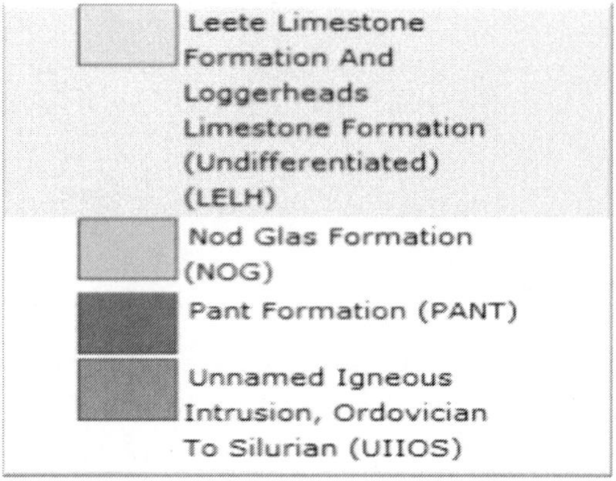

Figure 4.3: Geological Legend. (Source: Geology Roam. Digimap.edina.ac.uk/geologyroam, accessed 12/06/2013).

The map above showing Llanymynech Hill is highlighted with different colours that correspond to the geological legend indicating the different minerals found on and around the hill.

4.3 The Merits of Llanymynech Hill Limestone

Carboniferous limestone of Llanymynech Hill is a sedimentary rock made of calcium magnesium carbonate and calcite. The rocks are structured from shells and parts of sea creatures, encased in carbonate mud. Brachiopods, fossil and crinoids are components of Carboniferous limestone.

The limestone of the lower layers dominates Llanymynech Hill; it was generally considered to be the purest and so was exported in great quantities across England and Wales. Above can be seen the Leeke and Loggerheads limestone formation on and around Llanymynech Hill. The carboniferous limestone quarries on Llanymynech Hill were initially a source of lime which was burnt for agricultural purposes, to help improve the land. The limestone contains magnesium, which is good for the grassland producing rich meadowland for dairy cattle (Joan Zorn, Per Comm, 04/04/2013, Rhodes, 2008, Toghill, 2006, Ratcliffe 2004 and Gaffney, *et al,* 2007).

It is quite possible that there was mining on Llanymynech Hill before the Romans came in 43 AD. When the Romans reached Llanymynech they soon realised the potential of the minerals contained within Llanymynech

Hill. There is evidence of Roman copper mining activities throughout the 1st and 2nd centuries on Llanymynech Hill.

Although the primary mineral extracted was copper one of the four types of limestone available, a white porcellanious limestone, has been used in the 2nd century mosaic at Roman Wroxeter, which would have been transported down the River Severn to Wroxeter. This thus suggests that there was a settlement at that time for those working in the quarry on Llanymynech Hill (Joan Zorn, Per Comm, 04/04/2013, Ratcliffe, 2004, p3, Fewtrell, 1878 pp172- 232 and Gaffney, *et al,* 2007 p121).

4.4 Llanymynech, Wenlock Edge and Lilleshall Limestone for Agriculture Fertiliser

The quarries at Llanymynech mainly produced limestone for agricultural fertiliser; however other quarries such as Wenlock Edge Quarries also produced limestone for agriculture. Lilleshall Quarries had two types of limestone; the grey limestone which was unworkable could however also be used as agricultural fertiliser. The Wenlock Edge Quarries producing agriculture fertiliser were the Lea, Coates, Knowles and Hayes Quarries. In the 19th century 80% of the limestone produced at Lea Quarries was for agriculture. Although the production of agricultural limestone had been reduced from 80% to 20% since the beginning of the 19th century, in 1957, at Lea Quarries a crushing and milling plant was installed in order to enhance the production of crushed lime as an agricultural fertiliser. Some twenty years later in 1977, production at Lea Quarries was at 600,000 tons per year. The production figures for that year breakdown as: 80% aggregates, 10% crushed limestone for concrete, 8% agricultural fertiliser and 2% for fluxing stone (Williams, 1997 and Trinder, 2000).

Coates, Knowles and Hayes Quarries all produced limestone for agriculture; one of Coates' main customers being Lord Forester a local landlord. At the height of production Coates Quarry employed approximately 20 men, working day and night shifts, generating approximately 1,200 tons of limestone per week. The Knowles and Hayes Quarries were not on the same scale as Lea Quarries; however they did produce limestone for agriculture for a short time (Williams, 1997).

4.5 Quarry evolution: 19th – 20th century

An examination of old maps was undertaken to compare how Llanymynech Hill had changed from the 19th century to the 21st century. All the maps from 1880s to 1950s of Llanymynech and its surrounding area show six quarries on or around the Llanymynech Hill and one to the east of the hill. The main quarries showed on the maps of Llanymynech hill are the Welsh and English Quarries at the base of the hill. As can be seen below:

Figure 4.3: Llanymynech Quarries 1880. (Source: Digimap, www.ac.uk/ancientroam, accessed 22/05/2014).

Figure 4.4: Llanymynech Quarries 1890. (Source: Digimap, www.ac.uk/ancientroam, accessed 22/052014).

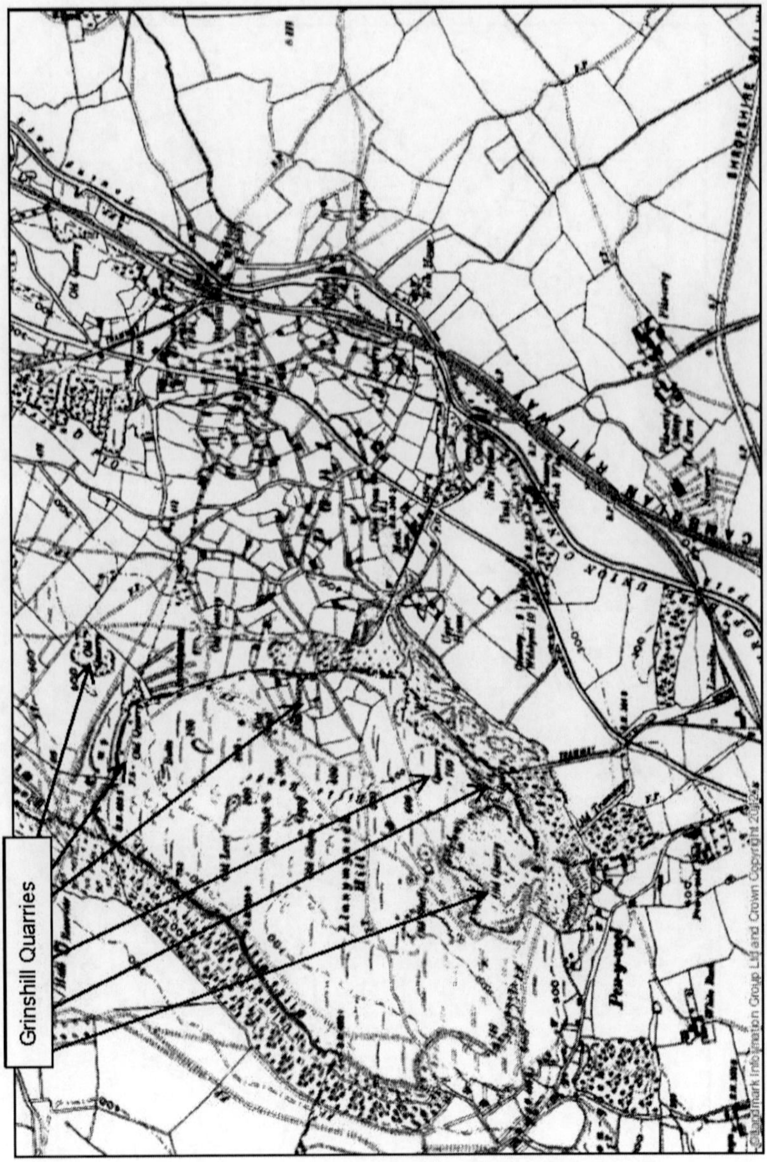

Figure 4.5: Llanymynech Quarries 1900. (Source: Digimap, www.ac.uk/ancientroam, accessed 19/05/2014).

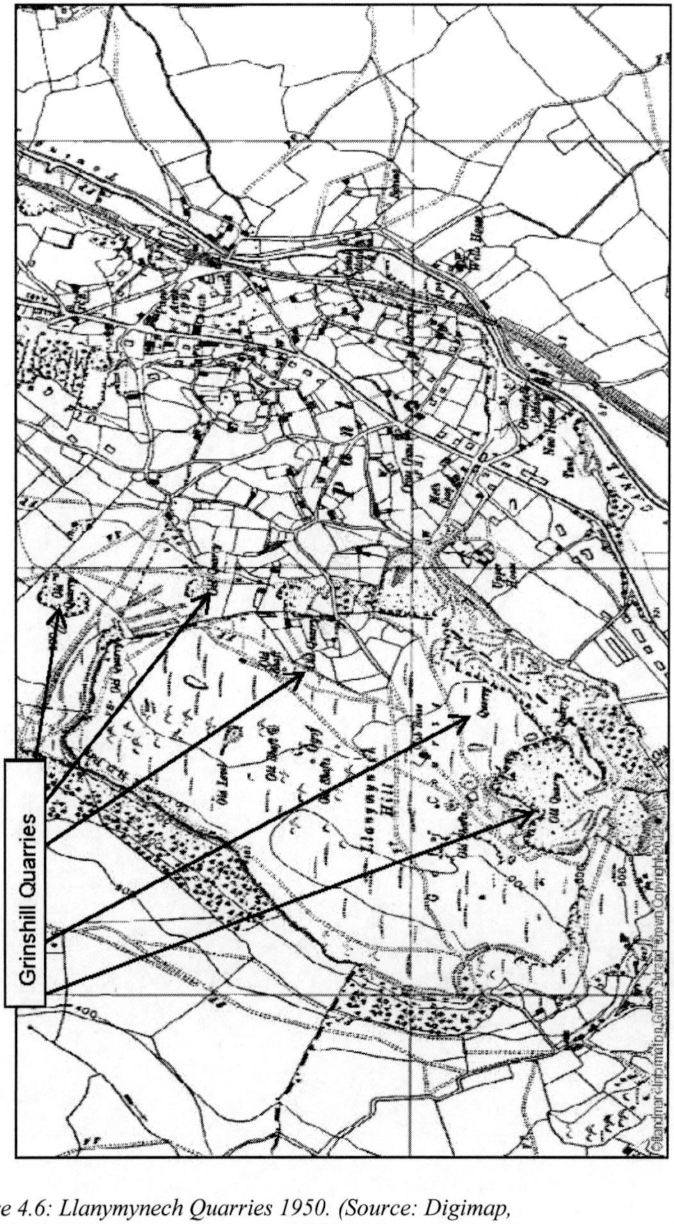

Figure 4.6: Llanymynech Quarries 1950. (Source: Digimap, www.ac.uk/ancientroam, accessed 19/05/2014).

4.6 The Llanymynech Industrial Period

With the start of the Industrial Revolution there was a need for a better transport system not only in Llanymynech, but across the whole of Britain, which would see the coming of turnpike roads, canals and railways. Overtime Llanymynech would benefit from all three innovations to the transportation system, which in turn would transform the whole limestone quarrying industry. Before all these new forms of transport came about, natural waterways such as the River Severn were initially used to transport minerals such as limestone. From river onwards, transport would have been by packhorse, or horse and cart (Trinder, 2000). Then around the mid-1750s a turnpike road, was built between Oswestry and Welshpool. Turnpike roads were planned to be laid as a straight line, necessitating the payment of a toll towards its maintenance, which was paid to the local group of entrepreneurs. Tolls were on a sliding scale dependant on the quantity being transported and amount of pack or draught horses using the road, so increasing the cost of the toll paid by the individual (Staffordshire Archives, Bradford Collection, document no, D1287/18/31 (K/164), accessed 16/04/2013 and Albert, 1972).

In the 1770's a second turnpike road was built via Knockin through Llanymynech linking it to the Shrewsbury road; an Act of Parliament in 1814 was passed for the maintenance and repair of the said roads. This allowed the quarries to transport greater quantities of lime byway of the turnpike roads for its use as fertiliser on the farm lands around Llanymynech and the surrounding counties (Zorn, J. Per. Comm. 04/04/2013). Also in winter the turnpike roads became very slow, so that horses were limited to what they could pull or carry (Albert, 1972 and Rhodes, 2008).

4.6.1 The Montgomery and Ellsemere Canal

A branch line, connecting to the Shropshire Union Canal, was built between 1794 and 1795 from Lower Frankton Canal to Llanymynech. The Montgomery Canal runs from Frankton Locks, a junction on the Shropshire Union Canal, to Newport in Montgomeryshire. Different maps show the canal reaching Llanymynech as being called the 'Ellesmere Canal' in 1818 and 1821, however an 1897 map shows it as the 'Shropshire Union Canal' (Harriet Devlin, Per Comm, 27/06/2013, Staffordshire Archives, Bradford Collection, document no, D1287/18/31 (K/164), accessed 16/04/2013 and Staffordshire Archives, Bradford Collection, document no, D1287/M/2246, Plan of Land Leased to Colonel West at Llanymynech, accessed 16/04/2013). The canal at Llanymynech was used to transport agricultural lime to Montgomeryshire. This increased the export of lime to approximately 56,000 tons per year and in order to process such large quantities, ninety-two lime-kilns were built along the 26 miles of this canal (Harriet Devlin, Per Comm, 27/06/2013).

In the 18th century all the quarrying took place on top of Llanymynech Hill, so consequently the excavated limestone was burnt in the kilns on the hill, then the quick-lime was transported down the hill by packhorse. Having the kilns near to the quarrying site had its compensations and difficulties, because once limestone has been processed into quick-lime it becomes volatile, producing great heat and burning anything it comes into contact with that is wet. The prevalence of wet weather around Llanymynech, taken together with the primitive state of the roads, increased the danger of transporting burnt lime, either on pannier ponies, or by horse and cart. So now it was only sensible to bring the quarried limestone down in its unprocessed state from the quarries to be conveyed by canal barge to the individual farmers' kilns along the canal bank. The limestone was conveyed along the Welshpool to Oswestry turnpike road, now the A483, from Llanymynech Hill and offloaded into barges docked on the northern side of the Ellesmere Canal which was close to this turnpike road. Even though the danger of quick-lime coming into contact with water remained, sometimes leaking through the hulls resulting in loss of life, cargo and barge, it was still transported by canal (Harriet Devlin, Per Comm, 27/06/2013 and Joan Zorn, Per Comm, 04/04/2013).

4.6.2 Llanymynech Hill Tramways and Inclines

The late 18th century onwards saw quarrying on Llanymynech Hill steadily increasing until around the middle of the 19th century when it reached its highest point, after which it started to decline. The increase in productivity at this time brought with it another problem, how to convey the enormous quantities of quarried limestone safely and efficiently down the hill from the quarries to the canal. This problem was overcome by the construction of a system of rail tracks. These tracks known as tramways, carried the limestone down the hill to the canal. The tramways conveyed the trucks, pushed by quarrymen or pulled by horses. This process was similarly used on Brown Clee Hill, Shropshire (Hewitt, 1991). Two tunnels were built so that the tramways could pass under the former turnpike road which ran between Welshpool and Oswestry; the original tunnels can still be seen today at the heritage site. The west tunnel was originally used by the Welsh, whereas the east tunnel was used by the English; they were both connected with their respective quarries by individual inclines. A photograph of the two tunnels can be seen below (Harriet Devlin, Per Comm, 27/06/2013, Tunnels under the A483 from Llanymynech www.discovershropshire.org.uk, accessed 20/06/2013 and Jones, 2004).

Figure 4.7: Llanymynech Heritage Area. (Source: Welsh and English Tunnels (Taken by Author 04/04/2013).

Records at the National Library of Wales indicate that in 1799 a Mr Smith desired to lease land from the Chirk Castle Estate for the purpose of quarrying limestone, with the suggestion that a tramway be constructed from the quarry to the canal at the estimated cost of £100 (National Library of Wales, Chirk Castle Estate, documents no, F8528 to F8573, accessed 05/06/2013). A proposal for the first tramway to be constructed at Llanymynech was put forward in December 1804 and completed in 1806. The tramway was constructed by Arthur Davis, Robert Cartwright and Richard Jebb, who also at that time, leased quarries from the Chirk Castle Estate and a Mr West (National Library of Wales, Chirk Castle Estate, document no, V/6050 and V6061, accessed 05/06/2013 and Ratcliffe, 2004). This tramway carried quarried limestone from quarries on land belonging to the Middleton family, owners of the Chirk Castle Estate, to the first canal wharf. At the top of the Welsh Incline were the Welsh Drum Brake Houses, the ruins of which can still be seen today. The incline carried a double rail track with a cross-over at the bottom, where the lines joined a single track to cross the Welshpool and Oswestry turnpike road, below which was a small passing bay. This single rail track continued to the side road next to the Ellesmere Canal, where it separated into two rail tracks running either side of the Welsh wharf, so that the limestone could be unloaded onto canal barges; a picture of this can be seen below indicating where the Welsh Incline was sited within the forest (Harriet Devlin, Per Comm, 27/06/2013, Tramways at Llanymynech, www.discovershropshire.org.uk, accessed 21/03/2013, Jones, 2004, p6 and Ratcliffe, 2004).

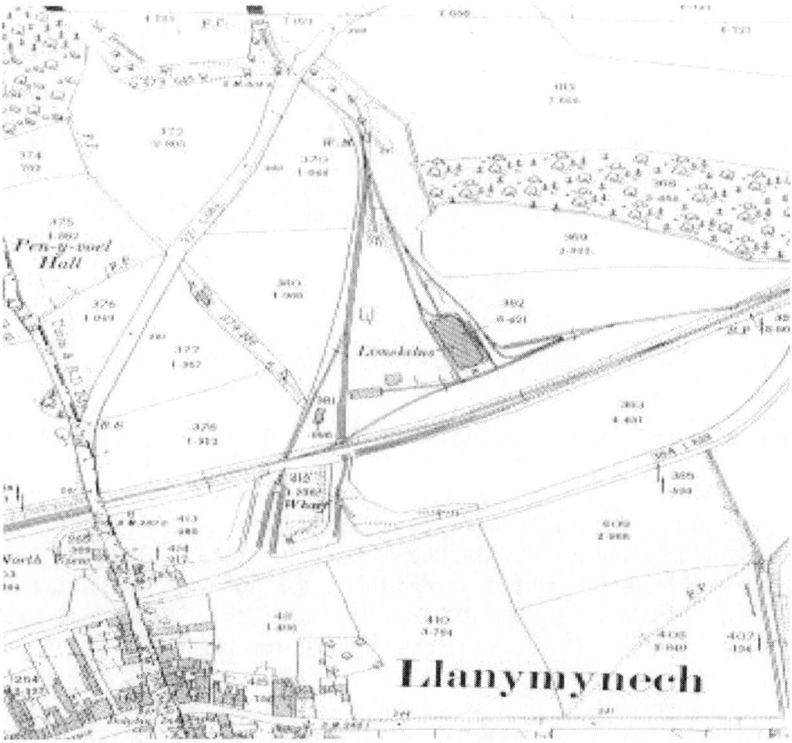

Llanymynech Tramways and Wharf

Figure 4.8: Llanymynech Heritage Areas, Llanymynech Tramways and Wharf. (Source: Clwyd-Powys Archaeological Trust-Projects, accessed 21/03/2013)

The track can still be seen within the incline route, especially where grooves have been cut into the stone; some to give a good grip for the horses and men, others cut by the wheels of wagons transporting the quarried stone down the incline to the canal. Similar inclines at Corbett Wood, Grinshill, still show grooves in the inclines today, which can be seen in Grinshill appendix 3. The Welsh Drum Brake House was probably constructed in the 1880s when a tunnel connecting the Welsh Quarry to the English Incline was built. In 1807, Thomas Yates put forward a proposal for a second tramway at Llanymynech to run from the Earl of Bradford's quarries to the canal. The planned tramway was never constructed as proposed, although a similar construction was laid down in about 1810 after Thomas Yates started to pay rent on leased land; this followed the original planned route (National Library of Wales, Chirk Castle Estate, document no, F6046, Plan of proposed tramway 1804, accessed 05/06/2013). The Ordnance Survey map of 1837 shows one of the inclines which was to become the main transport route, this could possibly have been the English Incline. As time went on, the route was steadily improved to make it more efficient and was still in use when the quarry closed in 1914 (Harriet Devlin, Per Comm, 27/06/2013 and Joan Zorn, Per Comm, 04/04/2013).

Figure 4.9: English Incline looking up from the base. (Source: Discovering Shropshire's History accessed 25/09/2013).

The English Incline, rising from about 1 in 4 to 1 in 5, leads to the English Drum Brake House, which was restored in early 2008. Pictures of The English Drum Brake House before and after restoration and a picture indicating where the Welsh Incline was sited within the forest are shown below (Joan Zorn, Per Comm, 04/04/2013 and www.discovershropshire.org.uk,_accessed 21/03/2013).

Figure 4.10: The English Drum House. (Source: Llanymynech Limeworks Heritage Project, accessed 16/05/2013).

Figure 4.11: English Drum House. (Source: Llanymynech Limeworks and Heritage Area accessed 06/08/2013).

Figure 4.12: The Welsh Incline looking up from the bottom. (Source: Clwyd/Powyis Archaeological Trust accessed 16/05/2013).

In 1766 the lime accounts of the Earl of Bradford (Bridgeman) show that there were expenditures totalling £190; this suggests that the Bradford Estate was active in limestone quarrying at this time. There are no other entries to do with quarrying in this ledger and no other ledgers could be located to confirm how long quarrying had continued after this date. So it is quite possible that the Bradford Estate relinquished its direct interests in limestone quarrying, leasing out mineral quarrying to others after this date (Jones, 2004 and Rhodes, 2008). An 1821 map is the earliest that indicates the land ownership where limestone quarrying was being quarried by the Chirk (green) and Bradford (red) Estates. The boundary between the Chirk and Bradford Estates demarcated the quarrying activities of both parties.

In the early 1820s there was a dispute between the Earl of Bradford, who owned the English Lime Works, and The Honourable Frederick West, who owned the Welsh Lime Works. The map of 1821 shows the limestone quarries owned by the Earl of Bradford and the Honourable Frederick West; the portion of land coloured red belonged to the Earl of Bradford, the major percentage, the portion coloured green were owned by the Honourable Frederick West. Both had tenants working on their individual lime works, who paid royalties to the landowners at so much per ton of limestone quarried. The yellow portion of the map indicates where the tenants of the Hon Frederick West had dumped waste stone and earth. Although the Hon Frederick West possessed the larger portion of workable land containing the majority of limestone, the greatest portion of land was in the ownership of the Earl of Bradford which separated Llanymynech Hill and the Ellesmere Canal. The tenants of the Hon Frederick West built a railway from his quarries across the land of the Earl of Bradford to the Ellesmere Canal, in order to obtain the shortest route. Mr Potter, the Earl of Bradford's agent, in his letter of 1821, put forward the consideration that the building of a similar railway for the benefit of the Earl's tenants could not be justified as only a reduced income could be gained from the limited amount of the remaining limestone. However, a later survey in 1822 calculated that in fact approximately three hundred thousand tons of limestone rock remained un-quarried (Staffordshire Archive, Bradford Collection, document no, D1287/13/86 (K/264) Potter letters, accessed 16/04/2013).

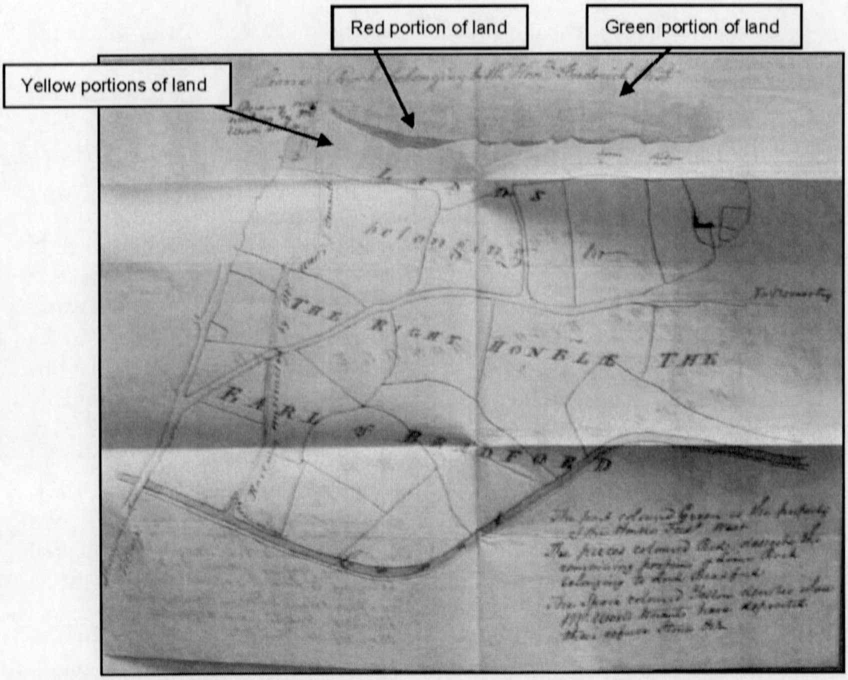

Figure 4.13: 1821 map showing the coloured portions of land. (Source: Staffordshire Archive, Bradford Collection, accessed 16/04/2013).

The Hon Frederick West's tenants charged the public for the use of the railway, 2/6, 2/3 and 2/- shillings per ton, reducing this to 1/9 per ton. The Earl of Bradford's tenants made a strong complaint about the charges saying that with their limited means, they were unable to compete with the Hon Frederick West's tenants without ruin to themselves. The tenants asked Mr Potter, the Earl of Bradford's agent, if the Earl would reduce their royalties, to 6d per ton, but as Mr Potter conceded to his employer "however it must be clear to your Lordship that by such a process no sort of relief could be afforded and your estate would sustain a loss". Another point put forward was that if the royalties were to be reduced, this action would only be followed by a further reduction in charges by the tenants of the Hon Frederick West. Mr Potter informed the Earl of Bradford that he had had several meetings with the tenants of the Hon Frederick West and expostulated at their unreasonable behaviour encouraging them to return to their former price of 2/- per ton, considering that this price would be acceptable to both parties, however it was not accepted. All of this could have been avoided if the Earl of Bradford had used his influence to stop the construction of the railway across his lands by the owners and occupiers of mines on the land leased from the Hon Frederick West (Staffordshire Archive, Bradford Collection, document no, D1287/13/86 (K/264) Potter letters, accessed 16/04/2013).

For several years, while constructing the railway, the tenants of the Hon Frederick West deposited enormous amounts of waste stone and earth on the Earl of Bradford's land. It was stated that it would take a considerable amount of time and cost, to remove the waste stone and earth. Mr Potter, the Earl of Bradford's agent, suggested that the Hon Frederick West be asked to remove the waste stone and earth. It was hoped that this measure might encourage the Hon Frederick West to act more reasonably and so allow the Earl's tenants a more favourable living. On the 26th May 1821 Mr Potter wrote to the Earl of Bradford stating these views. The Earl of Bradford replied by way of Lady Bradford on the 5th of June 1821, stating that he both approved of what had been done and what was proposed, but that he was averse to acting too quickly. Potter remarked that for a long time he had worked in the hope that the dispute might be resolved by coming to an amicable arrangement. The Earl's tenants continued to come to Mr Potter again and again declaring their inability to survive unless some relief was afford to them. It was apparent to Mr Potter that as the dispute had been going on for some time any further forbearance would be of no avail. So Mr Potter, the Earl of Bradford's agent, sent the offending parties a notice to remove the waste stone and earth; the agent's letter states that he hoped his actions met with his lordship's approval. It seems that the waste stone and earth was removed by the Hon Frederick West because the waste indicated on the 1821 map is not present on the 1822 map (Staffordshire Archive, Bradford Collection, document no, D1287/13/86 (K/264) Potter letters, accessed 16/04/2013).

Even with the removal of the waste, the Earl of Bradford's tenants continued to be harassed by the tenants of the Honourable Frederick West, because in addition in 1822 a toll gate had been erected across the Llanymynech to Oswestry road. A map accompanying the documents dated May 1822, shows the toll gate on the Llanymynech to Oswestry road, along which was the only route that the Earl of Bradford's tenants could travel with their wagons and horses from his lime works to the Ellsemere Canal. The distance from their quarries to the Ellesmere Canal being a matter of approximately 572 yards, whereas, the distance to the toll gate from the quarries is approximately 472 yards. The distance from the toll gate to the turnpike gate near the bridge at Llanymynech is approximately 100 yards, so for these 472 yards the tenants were charged £22:10:00 per year. Other parties using the road were only charged £10:10:00 per year, which had been a fixed rate for some years. The building of this toll gate brought a halt to the quarrying trade carried out by the Earl of Bradford's tenants. The Earl of Bradford's tenants were therefore forced to pay £22:10:00 a year until they relinquished their right to work the Earl's quarries, even when as they pointed out that by law, the authorised sum was just 3d for a horse crossing a toll gate on a turnpike road. The map below of 1822 indicates the lands belonging to the Earl of Bradford and the Honourable Frederick West. It also shows the "toll gate and the turnpike gate" which were at the centre of this dispute (Staffordshire Archive, Bradford Collection, document no, D1287/13/86 (K/264) Potter letters, accessed 16/04/2013).

The Earl of Bradford secured royalties for the lime extracted by his tenants from his lime works, which came to nearly £300 per year. In addition to this his tenants paid a toll on the quarried lime stone every time they passed through the toll gate onto the Oswestry Road, after which they then paid further charges at the turnpike gate near to Llanymynech Bridge. The tenants of the Honourable Frederick West paid no toll charges whatsoever because they transported their quarried limestone from the quarry face to their canal wharf byway of a railway constructed by Jubb and Cartwright. This caused the financial competition between the two parties to be unequal. To enable the Earl of Bradford's tenants to be on equal or perhaps similar terms, a proposal was put forward by Messrs Taylor and Baugh for the construction of a railway; the site of which is shown above on the 1822 map (Staffordshire Archive, Bradford Collection, document no, D1287/13/86 (K/264) Potter letters, accessed 16/04/2013).

Messrs Taylor and Baugh were two of the Earl of Bradford's most profitable tenants who proposed that they should give 10% of their profits and securities towards the construction of the railway by the Earl of Bradford. At that time royalties were paid at the rate of 5d per ton, and Taylor and Baugh proposed that they extract 800 tons of limestone each year for a term of 14 years. The basis of this proposal was made in the light of the fact that they had received an order for 10,000 tons of limestone, of which only 4,166 tons had been extracted. It was estimated that the cost of the construction of the proposed railway would amount to £803:00:00 with annual running cost of £80:06:00. An approximation of the quantity of limestone still remaining in the land belonging to the Earl of Bradford was carried out, exclusive of the land occupied by his tenants. The remaining limestone was estimated to come to approximately 300,000

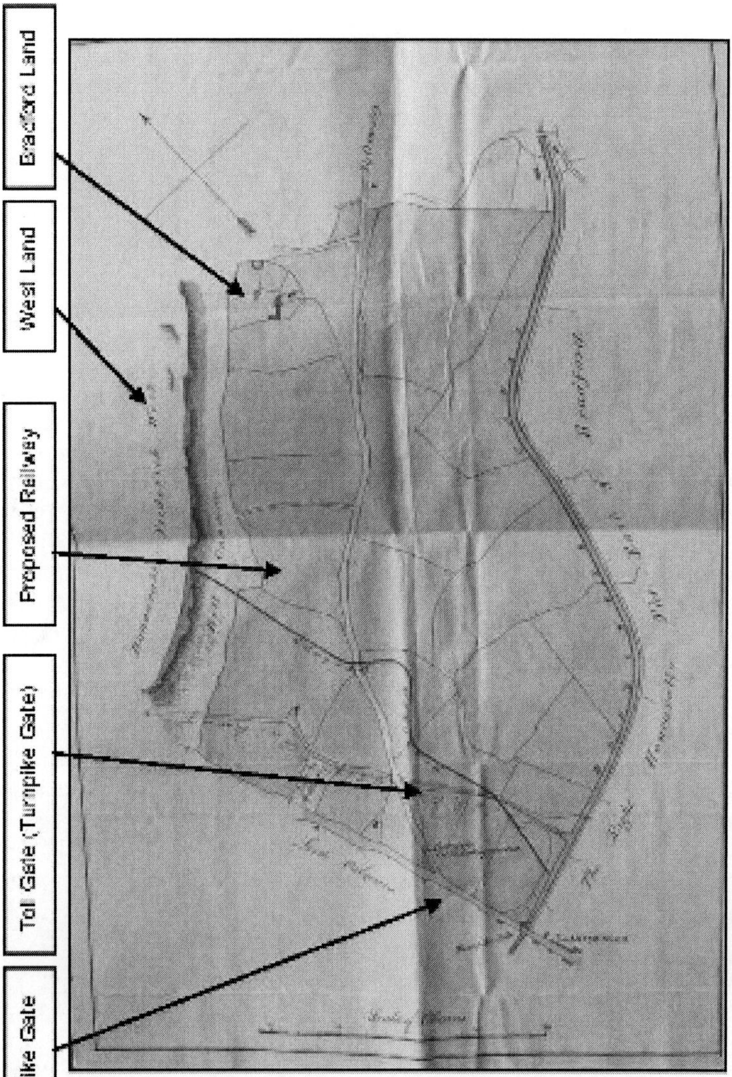

Figure 4.14: Potter letters. (Source: Staffordshire Archive, Bradford Collection, document no, D1287/13/86 (K/264) accessed 16/04/2013).

tons and it was suggested that this could therefore justify expenditure on the cost of building a railway, if approval was given by the Earl of Bradford. Mr Potter, the Earl's agent, suggested that this amount of money could be raised by the sale of oak timber on the Earl's estates at Blodwel and Brynor. Mr Potter stated that should the Earl of Bradford consent to the proposal, it would be easy to achieve the sale of the oak timber. He suggested that without these proposals being carried out, he could foresee greater difficulties arising between the two parties. It is not clear when the railway was finally constructed, however the proposed route can be seen on the 1822 map and it is possible that this railway had been constructed by the end of the 1820's.

Table 4.1 records the amount of limestone quarried by the tenants of the Earl of Bradford for the year 1821. It is informative to see the different amounts of limestone quarried by Owen. Owens, 4,278 tonnes, Taylor and Baugh 4,165 tonnes, Robert. Roberts. 2,242 tonnes, Hugh Pryce. 2,227 tonnes and Edward. Sockett. 1,883 tonnes. All the quarrymen had the same amount of time, being one year, however the amount of tonnage quarried is different. Taking each amount quarried from the highest total gives the difference quarried:113, 2,036, 2,051 and 2,365 tonnage. One can see from the calculations that there is quite a substantial difference, which brings one to ask why this was when all the quarrymen had the same time, one year. It is possible that some of the quarrymen worked longer hours and more days and therefore quarried a greater amount of limestone. Or should it be considered whether the sites quarried by the different quarrymen contained a higher quantity and a better quality of limestone. The weather for 1821 could also have played its part, and as I have said already, all the quarrymen had to work longer hours in the summer. These men were tenant owners of their individual quarry sites and it most likely would have worked all the time possible, because of the rent they would have had to pay for those quarry rights (Staffordshire Archive, Bradford Collection, document no, D1287/13/86 (K/264) Potter letters, accessed 16/04/2013 and Jones, 2004).

Table 4.1: Quantities of Limestone Obtained on the Earl of Bradford's Land, Llanymynech Hill by his Tenants, May 21st, 1821

Names of Tenants	Tons of Limestone 1821
Taylor and Baugh.	4165
Owen. Owens.	4278
Robert. Roberts.	2242
Hugh. Pryce.	2227
Edward. Sockett.	1883
Total	14795 at 5d £: S: D 308: 04: 07

(Source: Staffordshire Archive, Bradford Collection, document no, D1287/13/86 (K/264) Potter letters, accessed 16/04/2013).

There were numerous tramways at Llanymynech Lime works; the earliest had closed by 1838 and had been abandoned below the incline. A new track was constructed so that it connected with the two tramways directly west of the road. This had twin lines running underneath the road to a complex crossover, to the south where the western line followed a bend and then onto the Western Canal Wharf. The route the tramway took was known as the Old Rail Road; today only a narrow strip of land is left where a dwelling, known as the Rock Cottage, still stands (Shropshire Archives, Llanymynech Heritage Focus Group, Three-year Report 1998-2001, accessed 22/05/2013). In the 1830s Messrs Pickering and Co owned and ran the second tramway route, which went along the western side of the Welsh Quarry, down the incline and onto the canal. Messrs Baugh and Co owned and ran the third of the tramways at Llanymynech. Their incline went from The English Quarry towards Pant and down near to Bron Wylfa, near Llanymynech. The second and third inclines went through the tunnels under what is now the A483 (Rhodes, 2008 and Jones, 2004). The final two tramways went either side of the Tally House, where the waggons containing quarried stone would be recorded by the person manning the Tally House. After the stone had been recorded the waggons containing the quarried stone would proceed onwards to the Welsh and English Wharfs, and then loaded onto the canal barges (Harriet Devlin, Per Comm, 27/06/2013, Joan Zorn, Per Comm, 04/04/2013 and Rhodes, 2008).

4.7 Quarrying and Landowners on Llanymynech Hill

The limestone from Llanymynech was primarily used as agricultural fertiliser, but was sometimes used as lime mortar and as a flux in blast furnaces. Limestone quarrying at Llanymynech was at its height in the 1860s (Harriet Devlin, Per Comm, 27/06/2013, Bryan, 2006 and Rhodes, 2008). The principal landowners in and around Llanymynech Hill were the Middleton's of Chirk Castle and the Earl of Bradford, whose family name had been Bridgeman; both these parties owned the mineral rights for limestone quarrying. The Chirk Castle Estate quarried the southern side of Llanymynech Hill under a series of leases, which would become known as the Welsh Quarry. The estate of the Earl of Bradford quarried first the western side, and then later the eastern side of Llanymynech Hill, which would become known as the English Quarry. In the 1750s the Chirk Castle Estate seems to have been the first of the two estates to be linked to large scale quarrying on Llanymynech Hill (Jones, 2004 and Rhodes, 2008).

The portion of land between the two estates was never quarried but acted as the boundary, or a 'no-man's' land. It was apparent from the research undertaken that there was not a good working relationship between the Chirk and Bradford Estates; this can be seen in correspondence between the two parties (Staffordshire Archive, Bradford Collection, document no, D1287/18/31 (K/164), Griffiths letters and document no, D1287/13/86 (K/264) and Potter letters, accessed 16/04/2013).

Research undertaken at The National Library of Wales, shows records of the kilns worked at the Chirk Estate in 1775. Table 4.2 records the different kilns on the Chirk Estate and the tonnages of limestone processed at each individual kiln. The great kiln processed 8 tonnes at 5/- shilling per ton, the corner kiln 7 tonnes at 4/- shillings per ton, middle kiln 6 tonnes at 3s 6d per ton and the little kiln 4 tonnes at 2s 6d per ton. The locations and names of the kilns could have a possible meaning such as the 'great kiln' possibly named because of its size. Similarly, the names of the other kilns indicated position and size, down to the little kiln. So that the kilns sizes could simply mean that they could process more limestone and consequently charge more per tonne.

Table 4.3 records documented evidence found for the Chirk Estate for 1792, this provides a twelve month inventory of limestone quarried on Llanymynech Hill, together with an individual monthly account of limestone quarried. The records say "loads" yet this could also mean tonnage; it is interesting to note that the amount quarried each month is different. In January 1792, 92 tonnes, increased in February to 230 tonnes, a 138 tonne rise. The amount of limestone quarried continued to increase so that in March the amount of 422 tonnes was up by 208 tonnes. For some reason in April of the same year the limestone quarried fell to 104 tonnes, this could have been because of a possible wet month. In May 1792 the quarrying was up again to 272 tonnes. Over the following summer months excavation of limestone increased enormously. If one looks at the records of the summer months of 1792: June 1145 tonnes "load", July 1017 tonnes, and August 711 tonnes, the amount of limestone recorded is higher for each month, than for any other month of that year. This is evidence of the summer months being the most active in the quarrying industry at Llanymynech, as it was at Brown Clee Hill. However, there was no other option than to work for the quarry workers, for if they did not work, they would lose their jobs. The excavation of limestone started to fall rapidly from its height in June of 1145 to 327 tonnes in September, a decline in production of 818 tonnes of limestone. There is no reason why in October of 1792 the quarrying of limestone should pick up, it is a winter month, this could have just been a rare occasion. In the following two months of November and December the quarrying operations output fell to 104 and 128 tonnes respectively. This Chirk Estate record of the quarrying operations, not only shows what was happening at Llanymynech Hill in 1792, but also provides an insight into Shropshire quarry operatives. This is a unique record, because it shows that excavation increased in the summer months at Llanymynech and this would have been no different for Shropshire or for Wales. It also confirms that the winter months of the year were not the most productive for any quarry site. The quantity of quarried limestone would have been the same one suggests in the 19th century; the summer months being the most productive when compared to the winter (The National Library of Wales, Chirk Castle, document no, F10119, and F8528-8573 accessed 05/06/2013).

Table 4.2: Lime kilns and charges made for the use of each kiln in 1775

Kilns	Tons	Charges
Great Kiln	8	5/-
Corner Kiln	7	4/-
Middle Kiln	6	3/6
Little Kiln	4	2/6

(Source: The National Library of Wales, Chirk Castle, document no, F10119, accessed 05/06/2013).

Table 4.3: Chirk Castle Loads of Limestone Excavated 1792:

January, 1792:	92, Loads:
February	230
March	422
April	104
May	272
June	1145
July	1017
August	711
September	327
October	590
November	104
December	128

(Source: The National Library of Wales, Chirk Castle, document no, F8528-8573, accessed 05/06/2013).

A Chirk Castle document records lime burning carried out at Llanymynech Hill on the 30th December 1795. The table 4.4 shows payments for the month of December made for lime burning to Messrs: M. Baugh of £21:03:06, John Jukes, £18:12:06, James Witticase, £7:03:09, Hugh Pryce, 7s 6d and Edward, 5 shillings. There is quite a difference between the amount of the first two payments and the other three, particularly the last two of 7s 6d and 5 shillings. The time of the year being winter it is quite possible that there was limited amount of quarried limestone available at that time (The National Library of Wales, Chirk Castle, document no, 8545-8573, accessed 05/06/2013).

Table 4.4: Chirk Castle, 30th December 1795 work Lime Burning on Llanymynech Hill

Names:	Payments: £: S: D
John Jukes	18:12:06
M. Baugh	21: 03:06
James Witticase	7: 03: 09
Hugh Pryce	7: 06
Edward …….	5: 00

(Source: The National Library of Wales, Chirk Castle, document no, 8545-8573, accessed 05/06/2013).

In 1796 Richard Middleton of Chirk Castle died. As he was the last surviving male heir of the Middleton (Myddleton) family the estate was divided between his three sisters Charlotte, Maria and step-sister Harriet. An indenture of 1800 confirms this, showing the division as giving the sisters the rights to all mines, quarry seams, veins of limestone and beds of coal with slack and ironstone, which were at that time being worked or might be opened at some future time on any part of the lands (The National Library of Wales, Chirk Castle, document no, F8612, accessed 05/06/2013).

Additional papers for 20th December 1796 record that the lime being burnt on Llanymynech Hill was processed by nine people using different kilns. From 20th December 1796 to 22nd January 1797 the amount of cash paid on account was £55:0:0 for lime burning at this same site (The National Library of Wales, Chirk Castle, document no, F10119, and F8528-8573 accessed 05/06/2013). In 1798 and 1805 money was provided by the trustees of The Chirk Castle Estate to Mr John Evens for entertainment for the lime men. The first being on the 16th January 1798 for 19 dinners for the lime men at 2/- shillings per head totalling £1:10:0, 4 casks of ale at 6/- shillings totalling £2:2:0, pipes and tobacco, 2/- shillings and water also 2/- shillings the grand total for this amounted to £4: 4:0. Mr John Evens passed a receipt of the bill to a Mr W. Roberts. The second entertainment provided by John Evens on behalf of The Chirk Castle Estate was for 30 dinners for lime men at 9/- shillings per head totalling £3:0:0, 107 casks of ale at 8/- shillings totalling £3:11:4, tobacco 2/- shillings and water 2/6 shillings making a grand total of £6:15:10. This may show that the lime men may have been well respected by The Chirk Castle Estate for the trustees to have gone to such an expense (The National Library of Wales, Chirk Castle, document no, F8553 and F8571, accessed 05/06/2013). On July 29th 1805 a valuation was eventually carried out for Carreghove Manor, together with the lands and lime works at Llanymynech Hill which belonged to The Chirk Castle Estate. The valuation was carried out nine years after the death of Richard Middleton in 1796 for the purpose of dividing the estate between his three sisters, Charlotte, Maria and step-sister Harriet. The valuation was done in four lots. First, valuation was made of the land occupied by a Mr John Nightingale, next, the Carreghove Manor with undisclosed land and then, cottages and land in the occupation of several people. The final lot to be valued was the Llanymynech Lime Works, which at that time was occupied by Messrs. Arthur Davis, Cartwright and Jebb. The valuation for the lime works alone came to £882 and the grand total was £13,111 (The National Library of Wales, Chirk Castle, Valuation letter, document no F6057 accessed 05/06/2013).

It would seem that Messrs. Arthur Davis, Cartwright and Jebb would have been working the Chirk Castle limestone quarries, possibly until sometime in 1822, because the names of Jebb and Cartwright appear on a map of 1822 in the dispute between the Earl of Bradford and The Hon. Frederick West. After this period Messrs. Arthur Davis, Cartwright and Jebb appear in the rent book of the Earl of Bradford; however there is no further evidence to show that they were working for the Earl of Bradford. As the major portion of the limestone is sited on the Welsh side of Llanymynech Hill, it is questionable why they would change to the English side which had only a limited life span. The map accompanying the letter of 1822 from Mr Potter shows the names of Jebb and Cartwright written along the route of the railway track. It may have been that they were paying for the right to lay the railway track over the land, and not for quarrying rights. From 1810 to 1830 a Mr Thomas Yates is recorded as paying rent to the Bradford Estate for working limestone. However Mr Thomas Yates was only one of the Earl's tenants because correspondence between the Earl of Bradford and his agent, Mr Potter, refers to 'his Lordship's tenants' (Staffordshire Archive, Bradford Collection, document no, D1287/13/86 (K/264) Potter letters, accessed 16/04/2013, Jones, 2004 and Rhodes, 2008).

There seems to have been a dispute regarding the division of The Chirk Castle Estate between the three sisters Charlotte, Maria and step-sister Harriet. The dispute was settled by the Court of Chancery which directed that The Chirk Castle Estate should pass to the eldest sister, Charlotte Middleton, wife of Robert Biddulph of Ledbury, Herefordshire, who prefixed his surname with the additional name of 'Middleton'. In 1798 Maria Middleton married Frederick West and once the dispute concerning The Chirk Castle Estate was concluded, she was given a portion of the estate known as The Llanaron Dyffryn Ceiriog Estate, which included Llanymynech Hill. The other sister Harriet Middleton inherited Ruthin Castle and because Harriet never married, upon her death in 1848 the estate passed to the family of her late sister Maria, who had pre-deceased her in 1843. The West family lived at Culham Court, Berkshire, and their eldest son Frederick Richard West lived at Arnewood House near Lymington, Hampshire, before the family moved to Ruthin Castle after his Aunt Harriet's death (Ruthin Castle, www.denbighshirearchives, accessed 29/08/2013). In 1827 Ruthin Castle together with its lands had been part of The Chirk Castle Estate, significantly, now being in the ownership of Mr Frederick West. The land with the limestone quarries was leased out. It interesting to note in the same year his son, Frederick Richard West married Theresa John Cornwallis Whitby, only daughter and heir of Admiral

William Cornwallis and Mary Ann Whitby of Newlands. On his marriage he obtained the Newlands Manor Estate in Milford, Hampshire. On the death of The Hon. Frederick West in 1852 the lands and quarries passed to his eldest son Frederick Richard West, who thus inherited The Ruthin Castle Estate, from his aunt via his father, and The Llanaron Dyffryn Ceiriog Estate, through his mother. Frederick Richard West and his wife Theresa had two sons, the eldest Frederick Middleton West died in 1868 and never having married there were no heirs. The second son, William Cornwallis-West, and brother of Frederick Middleton West, therefore succeeded his brother and father in ownership of the estates. Records indicate that in the 1850s and 1860s Frederick Middleton West, senior, had had to sell some of his properties belonging to the Middleton and Chirk Castle Estates to pay for drinking and other debts (Ruthin Castle, www.denbigshirearchives, document no DD/RC/1073, accessed 29/08/2013).

It seems that after years of dispute between the Earl of Bradford and the Hon Frederick Middleton West, an agreement was eventually reached, but not until his son William Cornwallis-West inherited the family estates in 1868. An informal agreement had been made in 1873 between Orlando George Charles, the Earl of Bradford and William Cornwallis-West which allowed West to transport quarried limestone across the Earl's land. This agreement was ratified in 1878 with the granting of a 99 year lease allowing William Cornwallis-West to transport quarried limestone from his quarries to the canal and railway across the Earl of Bradford's land (Jones, 2004, and Ratcliffe, 2004).

In about 1827 The Hon. Frederick Middleton West leased his limestone quarries to Exuperious Pickering and Company for 31 years, replacing the former tenancies of Messrs. Arthur Davis, Cartwright and Jebb. In 1835 Exuperious Pickering and Company were recorded as paying rent of £15:5:0 for two railways and one stable from the Bradford Estate. In 1844 or 1845 Exuperious Pickering and Company was taken over by The Carreghofa Lime Rock Company, which may have however been a subsidiary company belonging to Mr Exuperious Pickering. In 1851 Frederick Richard West leased additional limestone quarries to a Mr T. E. Ward of the Lodge, Chirk; at the same time there was still a further 8 years to run on the quarries leased to Exuperious Pickering and Company. Longueville and Williams also leased quarries from the West Estate, as did a Mr Joseph Needham, whose quarry lease was extended for a further two years. In 1856 leases were granted to Messrs. Benjamin Manning and John Dicks to carry out limestone quarrying. In the Ruthin Estate papers, recently released on-line, there is no reference to any trustees being appointed on the death of Frederick West in 1852, yet certain documents quote Messrs. Longueville and Williams as being trustees (Ruthin Castle, www.denbigshirearchives, document no DD/RC, accessed 29/08/2013, Rhodes, 2008, Jones, 2004).

Cottages in the village of Llanymynech owned by the Earl of Bradford were rented to the quarrymen; however these cottages were tied to the job, so that the company had a hold over the quarrymen's employment. The quarrymen had to fulfil specific quotas of worked limestone in order to keep a roof over their family's head, which meant long hours of extra shifts being worked in the summer (Joan Zorn, Per Comm, 04/04/2013 and Staffordshire Archive, Bradford Collection, document no, D1287/13/86 (K/264) Potter letters, accessed 16/04/2013).

In the 1860s the Earl of Bradford's lands at Llanymynech Hill were leased to the Carreghofa Lime Rock Company, although by this time some of the limestone quarries owned by the Earl of Bradford had been worked out and closed (Harriet Devlin, Per Comm, 27/06/2013 and Thomas Savin, www.Llanymynech.Llimestonequarries, accessed 09/06/2013). In the early 1860s a Mr Thomas Savin and Company, who had been very involved in the Cambrian Railway, acquired leases for both the Welsh and English Quarries. This would see an increase in limestone extraction and therefore Thomas Savin needed to extend the canal wharfage. This increase was to accommodate the extra limestone being produced and to provide a better and larger transport system for the limestone going out and the coal coming in to Llanymynech. In the mid-1860s Mr Thomas Savin became subject to financial problems created by the fact that he was involved in many different companies, not just in Britain but throughout the world. Creditors, including contractors, wanted their money, so on the 5th February 1866 Mr Thomas Savin was forced to declare himself bankrupt. Having declared himself insolvent, it was found that he owed over £2 million. All Mr Thomas Savin shares in different companies were taken from him to pay his debts; after all this he was left with four small companies, one being the Llanymynech Limestone Quarries. Following this the lime works at Llanymynech were re-named as The Porthywaen and Llanymynech Lime and Limestone Works, with their offices based at Cambrian Buildings, Oswestry (Harriet Devlin, Per Comm, 27/06/2013 and Thomas Savin. www.Llanymynech.Llimestonequarries, accessed 09/06/ 2013).

A letter sent from Mr Griffiths, an agent for the Earl of Bradford, received in the Earl's offices on the 4th June 1874 refers to the fact that the Earl's limestone was nearly worked out, with the exception of the land beneath the spoil heaps. It also confirms that Mr West's limestone works were much larger and still in the process of extracting limestone. Mr

Griffiths also asked the Earl if he recalled a proposal made to Mr West, through his agents, earlier in the year, to which he had had no response. This proposal referred to the possible construction of a future railway by Mr West. The agent pointed out that a Mr Savin and the Cambrian Railway had tried to obtain an Act of Parliament for a railway line to be constructed earlier. This caused the Earl of Bradford enormous expense in defence of his rights (Staffordshire Archive, Bradford Collection, document no, D1287/18/31 (K/164), accessed 16/04/2013). Special authorization for the drawing up of a lease was asked for from the Earl of Bradford. The lease was to include provision made with regards to what could, or could not be done, with the extant spoil banks, tramways, canal wharfs and canal basin. The agents of Mr West were keen to come to an agreement, however the agent of the Earl of Bradford wanted to make them wait, stating that they had already had plenty of time. The agent suggested that Mr West should be made to wait until their proposal had been thoroughly discussed with the Earl of Bradford in order to settle the matter (Staffordshire Archive, Bradford Collection, document no, D1287/18/31 (K/164) Griffiths' letters accessed 16/04/2013).

The Porthywaen Limestone Company closed the Llanymynech Lime works on the 14th August 1914, just after the outbreak of the First World War. An interview with a Mr Robert Cecil Corfield in 1983 records that, at the time of it closure, the lime works employed about 30 men. Before the lime works had closed, Mr Robert Cecil Corfield had been employed as an office boy. At that time it was still considered to be a very productive limestone quarry. He remarked that, just as it had been throughout the 19th century, quarrying limestone was still very hard work. Similarly, as in the 19th century, the weather played its part in the activities of the quarry, as when the weather was bad the quarrymen could not work and if they did not work, then they did not get paid. The closure of the lime works at Llanymynech was a great loss to the village; however most of the men found employment in the Blodwel and Criggion Limestone Quarries. Another problem for the lime works at Llanymynech in the early 20th century had been that the demand for quick-lime had begun to decline, as new chemical fertilisers had been developed. At the same time, lime as a mortar ceased being used as a building material (Rhodes, 2008 and Shropshire Archives, Interviews with Mr Robert. Cecil Cornfield, document no's, F L 56 and C 56, accessed 22/05/2013).

In 1914 production at Llanymynech Lime Works was greatly reduced, and because of the immense processing capability of the Warren Hoffmann Kiln, constructed in 1899, its size made it no longer financially viable to operate and it was closed. Small scale production continued for approximately another thirty years, until operations finally ceased in the 1930s. It is suggested in "Rhodes, 2008, p 68" that the First World War was not the reason for the closure of both the Llanymynech Lime Works and The Warren Hoffmann Kiln. However, from research already undertaken into the Lilleshall Company and Wenlock Edge Quarries, the First World War did greatly affect the quarrying industry during and after that time. This was because of the widespread shortage of manual labour, so that consequently productivity suffered at every quarry. The same shortage of manual labour before and after the war was nationwide and would have affected every industry, so that this would have been equally applicable to the Llanymynech Lime Works (Trinder, 2000).

4.8 Conservation and Restoration of Buildings, the Canal and Wharves Associated with the Quarrying Industry at The Llanymynech Lime Works

After the closures of Llanymynech Lime Works, at the same site, the Tally House, tracks and other buildings fell into disrepair and were reclaimed by nature. Talks took place and after many discussions a grant of £900,000 was allocated for the restoration and conservation of The Llanymynech Lime Works Site. The grant was obtained from The Heritage Lottery Fund, English Heritage, The Aggregates Levy Sustainability Fund, Shropshire County Council, Oswestry Borough Council, British Waterways and Llanymynech and Pant Parish Council. The grant was made for the restoration and stabilisation of the English Brake Drum House, inclined planes, kilns, stables, Tally House, canal, wharves and general pathways. Without this grant The Llanymynech Lime Works would not be what it is today and would have been lost to time. The project started in March 2006 and ran for a period of three years until February 2009. It is noteworthy that all these different associations worked together for the purpose of the restoration and conservation of such a major site. This is one of the best preserved sites associated with limestone quarrying not only in Wales, but also in England, as it gives a broad insight into past centuries of the quarrying industry (Joan Zorn, Per Comm, 04/04/2013 and Ratcliffe, 2004).

4.8.1 Welsh and English Brake Drum Houses

The Welsh and English Brake Drum Houses were used to control the movement of the trucks carrying the quarried limestone up and down the steep incline, from the individual quarries. The cable which was wound around the brake drum connected to the full truck at the top and the empty truck at the bottom of the incline. This would enable the empty truck to be pulled back up by the force of the filled truck going down. There were two parallel sets of tramway tracks one going down and the other up, so that the empty and filled trucks never met. The English Brake Drum House was worked by a Brakeman, this was frequently a young boy; it is possible that the same situation existed in the three Welsh Brake Drum Houses. Nevertheless; the safety measures that needed to be in place must have been similar in both Welsh and English Brake Drum Houses. Natural wear and tear on the cables such as fraying of cable wires, together with human error could result in the derailing of the trucks. Some inclines had safety features, for example check points part-way down the track, usually in the middle. On each side of the tracks were banks of soft sand, into which runaway trucks could be diverted. The brakeman needed to be alert to the possibility of a runaway truck, so that he could throw the switch to avoid any catastrophic crash. Should this have occurred while the Drum House was controlled by a young person, this would have been a greater responsibility than such young shoulders should bear. One suggests that this responsibility should have been given to someone considerably older and more experienced (Welsh and English Brake Drum Houses, www.discovershropshire.org.uk, accessed 20/06/2013 and Photographs and Notes taken on site visit to Llanymynech Lime Works Heritage Area by Galloway, 04/04/2013).

Before 2008, The English Brake Drum House was in ruins. The two stone walls, a cross beam and the iron band of the brake drum, were all that were left of the former building which was covered in bushes. Sometime before 2008 work was started to remove all the bushes from the stone ruins of The English Brake Drum House, repairs having been carried out to the stone work. After this initial reconstruction, in April 2008 a replica brake drum weighing over two tons was installed into the former ruins (Replica English Brake Drum, www.discovershropshire.org.uk, accessed 06/08/2013).

The preservation and conservation of the three Welsh Brake Drum Houses took place in 2011; the work was undertaken by L. J Preece Builders. It became obvious after the clearance of the bushes and rubble from these Brake Drum Houses that the buildings needed to be stabilised to preserve them for the future. For years The Welsh Brake Drum Houses had been empty and therefore time had taken its toll on the walls. Some of the walls had fallen down and needed to be re-built while the existing walls needed re-pointing. Finance was obtained from: Powys Council, CADW which is the Welsh Governments Historic Environment Service for the protection of the historic environment in Wales, Shropshire County Council, Pant and Llanymynech Parish Council and the former Oswestry Borough Council. Architect Mr Tim Ratcliffe managed this restoration and conservation project of The Welsh Brake Drum Houses (Joan Zorn, Per Comm, 04/04/2013 and Hankinson, *et al,* 2011).

4.8.2 Tally House

The Tally House was built with a weigh-bridge in the 1870s for the tallyman to weigh each truck load of quarried limestone brought down from the quarry face. The House Ledger for 22nd of October 1910 records and provides evidence for quarry excavations carried out by gangs of quarry workers, who were employed on a piece-rate system. The small teams of quarrymen were led by a foreman or team leader. The team's payment was based on how much quarried limestone they could quarry and deliver each working day. Every truck had its own "tally"; a brass token marked the specific number which only applied to the foreman, or leader, of each team. The tallyman after weighing the gang's quarried limestone would then direct the trucks of quarried limestone down to the canal, kilns or railway. The tallyman with this special number token given to each gang, was able to keep a record in his ledger of the individual team's production, daily and weekly. If one examines the ledger recording the limestone quarried by each gang, there is quite a difference. Gang one seem to quarry more limestone on Wednesday (21 tonnes), Thursday (28 tonnes and 16 cwt) and Friday (21 tonne). Gang two's highest day of production was Friday (20 tonnes and 3 cwt quarried). Gang three quarried 33 tonnes on Wednesday and 2 cwt. The other gangs four, five, six, seven and eight seem have quarried around the same amount each day within a few tonnes and cwt. The start of the week for each gang seems to be slow, however again gangs one and two are the leading teams, only over taken on Saturday by gang three and four by the amount of limestone quarried. However it would have been no good being the gang quarrying the greatest amount of limestone, if a day's work was missed. Each gang's day's absence from quarrying has been recorded during one week: gang two on Tuesday, gang five on Wednesday and gang six on Thursday. This is only one week's ledger of each team's work, however if this should represent what was happening every week of the month, it would soon add up to a large amount of time lost. I would suggest Chirk Castle would not have had any objections because if the gangs did not work they were not paid. One would have thought that working on the piece rate system would have ensured that the gangs of men would have worked as many hours and days as possible. The weekly total column shows how each gang performed by the end of the week: gang one, 107 tonnes and 1cwt; gang two, 52 tonnes and 18 cwt; gang three, 108 tonnes and 10 cwt; gang four, 94 tonnes; gang five, 42 tonnes and 18 cwt; gang six 42 tonnes and 13 cwt; gang seven, 68 tonnes and 8 cwt; and the last gang eight, 32 tonnes and 8 cwt. The gangs that were producing the greater quantity of quarried limestone were gangs one, and three followed by four, and this was in the winter months. What would have been the team's limestone production in the summer months? What does this say about what we know about the fact that there was less work in the winter months? However it should be taken into account that as the gangs was employed on a piece rate system, the gangs would probably have worked as many hours as possible. The Tally House Ledger is a unique record giving evidence of one week's work together with the amount of limestone quarried at Llanymynech (Joan Zorn, Per Comm, 04/04/2013 and Tally House, Llanymynech, www.discovershropshire.org.uk, accessed 06/08/2013).

Table 4.5: Tally House Ledger for the 22nd October, 1910

Tally House Ledger for the 22nd October, 1910. Weight of Limestone per Team.														
	Monday		Tuesday		Wednesday		Thursday		Friday		Saturday		Weekly Total	
	Tons	Cwt.	Tons	Cwt.	Tons	Cwt.	Tons	Cwt.	Tons	Cwt.	Tons	Cwt.	Tons	Cwt.
Gang 1	11	17	12	00	21	00	28	16	21	08	12	00	107	01
Gang 2	12	13	-	-	2	13	5	01	20	08	12	3	52	18
Gang 3	5	16	10	12	33	02	33	01	9	18	16	01	108	10
Gang 4	9	17	30	15	10	03	14	11	13	03	15	06	94	00
Gang 5	7	12	10	02	-	-	7	11	10	04	7	09	42	18
Gang 6	7	09	10	01	2	12	-	-	13	01	9	10	42	13
Gang 7	9	15	5	15	15	11	16	04	16	00	5	04	68	09
Gang 8	3	19	2	10	7	15	7	13	7	19	2	12	32	08
Daily Total	68	18	81	15	93	01	112	17	112	01	80	05	548	17

(Source: www.Llanymynechlimestonequarries/4Ftons.tallyhouse, accessed 27/09/2013).

Table 4.6: Tally House Ledger showing destination of quarried limestone

	Destination of Quarried Limestone, decided by tallyman					
	Canal		Kilns		Railway	
	Tons	Cwt.	Tons	Cwt.	Tons	Cwt.
Monday	20	02	40	15	8	01
Tuesday	10	06	57	10	13	18
Wednesday	-	-	46	06	46	15
Thursday	21	07	65	06	26	04
Friday	18	05	59	11	34	05
Saturday	-	-	30	15	49	09
Total	70	00	300	03	178	12

(Source: www.Llanymynechlimestonequarries/4Ftons.tallyhouse, accessed 27/09/2013).

Table 4.6, above, from the Tally House Ledger, gives no date for the limestone recorded in the tallyman's ledger. The document shows whether the limestone was destined to the canal, kilns, or the railway. Limestone that had been transported and processed in the kilns was then moved to take route either by canal or railway. The canal and the railway was the route processed limestone was transported and the kilns where it is burned before it went to the canal or railway. One could possibly suggest that if the canals were still being used, it must have been around the time when the railways were becoming more prominent. By the middle of the 19th century there was a network of railways. We know that the Tally House was built in the 1870s, so this table must have been created sometime between then and 1910. Examination of the table clearly shows that the railway transported a greater tonnage than the canal. This is quite clear from the amount of limestone being moved each day, the total columns for the canal is 70 tonnes and for the railway is 178 tonnes and 12 cwt, so, there is a nearly 100 tonnes difference. The table shows that the canal was not used on two days, whereas the railway was used every day. It also records each day the limestone was moved by the canal and railway. The canal starts on Monday transport with more limestone, yet this declines by the end of the week with the railway moving 49 tonnes and 9cwt on Saturday. The kiln column shows that a total of 300 tonnes 3cwt had been processed at the week's end. The kilns processed more limestone on Thursday, 65 tonnes and 6 cwt, Friday, 59 tonnes and 11 cwt and Saturday, 30tones and 12 cwt. This is a unique record which gives insight into the working life of the Tally House and Llanymynech Hill Quarries

(Tally House, www.Llanymynechlimestonequarries/4Ftons.tallyhouse, accessed 27/09/2013).

From the 1970s to 2006 the Tally House was covered in ivy and had to be fenced off as the building was becoming unstable and therefore a danger to the public. In March 2007 work started on the Tally House, the ivy was removed and the brickwork made safe and ready for restoration. By the end of June that year, some of the brickwork had been re-laid and the building was beginning to regain its shape. On my site visit to Llanymynech Lime Works Heritage Area on the 4th April, 2013, the Tally House had been restored in part, but was without its roof. From the restoration work carried out one could imagine the total size and shape of the building. Outside the Tally House is a truck full of limestone standing on a steel plate representing the weigh-bridge. There is a 2-D sculpture representing a tallyman weighing the full truck of stone (Tally House, Llanymynech,www.discovershropshire.org.uk, accessed 06/08/2013).

4.8.3 Warren Hoffmann Kiln and the Draw Kilns

There was an enormous market for quick-lime, both as fertiliser, in heavy industry and construction. The kilns would fall into two categories, ones near to the quarry face and others nearer to the canal and railway. The first category were three bank draw kilns which were located south-east of the present Wildlife Reserve. There were also two enormous draw kilns west of The Warren Hoffmann Kiln, which as with the other draw kilns, had a single draw-hole and were always in continuous operation. With an increase in demand for limestone there was a need for larger kilns to process the mineral into quick-lime. In 1899 The Warren Hoffmann Kiln was constructed which would lead to the demise of all five kilns by the 1900s. The Warren Hoffmann Kiln itself had only a short life of some fifteen years as the quarry closed in 1914 (Harriet Devlin, Per Comm, 27/06/2013, Jones, 2004 and Rhodes, 2008).

4.8.4 The men working The Warren Hoffmann Kiln

The Controller of The Warren Hoffmann Kiln was a man with overall responsibility based on knowledge gained over time for this highly skilled job; for instance he would be able to judge how the lime was burning simply by the colour of its flame. He also controlled the speed and direction of the burn by the amount of coal used. The coal was wheel-barrowed in by other workers across the top of the kiln and stacked around the stoke holes, of which there were three hundred and sixty-three. The

Controller could also regulate the draught dampers of the kiln by sliding them horizontally across the flue holes at the base of the kiln. All these flues were connected to the single chimney, which drew the fire up and released the smoke (The Warren Hoffmann Kiln, Llanymynech, www.discovershropshire.org.uk, accessed 06/08/2013).

One of the worst jobs must have been collecting and loading up the quick-lime into trucks after it had been processed. The men employed to do this job wore caps and scarves to help protect their skin and lungs, regardless of the enormous heat in the kiln. Their sweat and the lime dust mixed together made a very dangerous burning blend. This blend must have damaged their health, in particular from breathing-in the lime dust which would cause the burning of their lungs; this must have shortened their lives considerably (The Warren Hoffmann Kiln, Llanymynech, www.discover shropshire.org.uk, accessed 06/08/2013). Before the process of burning the limestone could begin, the kiln had to be loaded; this was done by skilled men known as "Stone Packers". These men would build walls of limestone; the quarried limestone was brought in on wheelbarrows and trucks. Although it must have been a slightly healthier job than loading up the limestone after it had been processed, there would still have been some dust about, including on the floor of the kiln chambers. While some jobs in the kiln were better than others, overall it was still not a healthy place to work in the 19th century (The Warren Hoffmann Kiln, Llanymynech, www.discover shropshire.org.uk, accessed 06/08/2013).

4.8.5 Restoration of The Warren Hoffmann Kiln

After the closure of The Llanymynech Lime Works in 1914, The Warren Hoffman Kiln was used as a cattle shed for some years. As with the other buildings on the site, over time, The Warren Hoffmann Kiln began to fall into disrepair and natural elements such as bushes, trees and the weather took their toll on the building (The Warren Hoffmann Kiln, Llanymynech, www.discovershropshire.org.uk, accessed 06/08/2013).

In the late 1970s Shropshire County Council purchased The Warren Hoffmann Kiln site from the Earl of Bradford's Estate. After the purchase of the site some basic conservation work was carried out to stop any further deterioration. One of the first things to be done was the placing of a waterproof membrane on the top of the kiln to make it safe from the natural elements; yet the waterproof membrane proved to be ineffective and may even have caused more damage in the long term. The chimney was also in urgent need of repair, so steel bands were used to retain it. The exterior brickwork of the chimney needed repointing and the replacement of some bricks. Following my site visit on the 4th of April 2013, I could see where brickwork had been replaced and the chimney wall pointed up. Another major factor which needed to be taken into consideration at the time of restoration, was that The Warren Hoffmann Kiln was now occupied by bats, whose roosting quarters are protected by law (Warren Hoffmann Kiln, Llanymynech, www.discovershropshire.org.uk, accessed 06/08/2013 and Photographs and Notes taken on site visit to Llanymynech Lime Works Heritage Area by Galloway, 04/04/2013).

Following the clearance, in 2003, of the bushes, trees and other rubbish around The Warren Hoffmann Kiln site, the depredations caused to the kiln and its surrounding area could be analysed. One of the first important things to be done was to make the roof water proof. In early 2007, after the removal of the temporary membrane which had been covering

the roof of the kiln, a steel frame was erected around and over the kiln. It was planned that in time a framework would be built over the kiln and this would be covered with sheets of weather-excluding plastic material. While the frame was being constructed, restoration to the external and internal brick walls of the kiln was underway. Where the internal roof of the kiln was sagging a series of five drystone walls was erected to support the roof. Internally, the walls were additionally supported with concrete blocks to act as reinforcement, while the dry internal walls were covered with a wire netting to protect people from the hazard of falling rubble. As well as drystone walls, timber supports were used to underpin an area of roof where sagging had happened. An open day was held in April 2008 so that the extent of the restoration work to the kiln could now be seen. A sculpture of a workman pouring coal down the stoke-holes has been placed on the kiln roof and three further sculptures of stone-packers have been placed outside the kiln, as if taking a break from work. On my site visit on the 4th of April 2013, I was able to walk around and stand in the restored Warren Hoffmann Kiln and review the entirety of the works. Additionally, these extensive works include the turntables that would have made it easier to manoeuvre trucks loaded with processed limestone from the kiln to the main railway line (The Warren Hoffman Kiln Restoration, www.LlanymynechLimestoneQuarries, accessed 27/09/2013, Harriet Devlin, Per Comm, 27/06/2013 and Photographs and Notes taken on site visit to Llanymynech Lime Works Heritage Area by Galloway, 04/04/2013).

4.8.6 Canal, Wharves, Stables and general pathways

Renovations to the former stable blocks were completed in December 2007; these had accommodated the horses, ponies and donkeys that moved the trucks of limestone onto the tramways. These stable blocks have been adapted so that they can now be used as an Education Centre open for visitors as well as for the Friends of Llanymynech Lime Works for their meetings Visitors can see a model of the former Llanymynech Lime Works, as well as learning about aspects of its history from information boards, which provide diagrams of the restoration process of all the buildings on the site. These information boards include facts relating the involvement of Mr Thomas Savin, an earlier proprietor, and Mr John Roberts, Forman, checking and grading the limestone blocks. It is an excellent building providing an ideal place from which to start before walking around the whole site. There is another Visitor Centre at the side of the canal, which had originally been the stables belonging to The Sun Inn, since demolished in the 1950s; this is now where visitors can purchase refreshments (Stable Block Restoration, www.Llanymynech LimestoneQuarries, accessed 27/09/2013, Hannaford, *et al,* 2008, and Photographs and Notes taken on site visit to Llanymynech Lime Works Heritage Area by Galloway, 04/04/2013).

On my site visit and interview with Mrs Joan Zorn, it was clear to see where the restoration work had been carried out to the canal, wharves and the footpaths. Work was still on-going to one of the wharfs, but lots of vegetation had been cleared away to give a better understanding of both the canal and the wharves. It was still clear to see that there was work in progress clearing vegetation in some places, yet overall the footpaths were in an excellent condition (Joan Zorn, Per Comm, 04/04/2013, Hannaford, *et al,* 2008, and Photographs and Notes taken on site visit to Llanymynech Lime Works Heritage Area by Galloway, 04/04/2013).

4.9 Archaeology and Industrial Archaeology

The late 18th century saw the start of the Agricultural Revolution which drove the need for more quarrying of limestone and thus an expansion in the quarrying industry followed. The growing demand for more Llanymynech Limestone to be processed into quick-lime for use as an agricultural fertiliser increased the need for a better transport system. This lead to the building of The Ellesmere Canal, which reached Llanymynech in 1796. With the coming of the canal The Llanymynech Lime Works, took the opportunity to expand and explore new industrial markets. This expansion was complimented by the construction of new systems of tramways and inclines at the quarry site, to convey the goods to a newly built distribution point at the canal. In order for the limestone industry to continue and develop, it had to constantly adapt to obtain new markets. So with the coming of the Llanfyllin Railway in 1863, The Llanymynech Lime Works did precisely that, it re-adapted and grasped the new markets (Ratcliffe, 2004 and Rhodes, 2008). As the end of the 19th century approached the limestone industry at The Llanymynech Lime Works began to decline. This was due to more economically viable alternative fertilisers and building materials becoming available from other sources. Even with the construction of The Warren Hoffmann Kiln and its greater output capacity, the 'writing was on the wall' and the Llanymynech Lime works closed in 1914 (Ratcliffe, 2004 and Rhodes, 2008).

The buildings, whether belonging to the English or Welsh Quarries, play an enormously important part in the industrial archaeological heritage for both England and Wales. Within The Llanymynech Lime Works a number of buildings were granted listed building status in March and October 1987 and therefore any work that needs to be carried out on these structures now needs to gain a Listed Building Consent. The buildings protected are: The Warren Hoffmann Kiln and its Chimney Grade II and the two lime kilns to the west, also being Grade II (The National Heritage of List for England, www.LlanymynectLimestonequarries/1021412, accessed 15/10/2013, Jones, 2004 and Ratcliffe, 2004).

Although these are the only buildings listed on the site at this time, I would suggest that the other buildings and structures, such as the Drum Houses, Incline Planes, the Tunnel under the A483, and The Tally House are of equal importance because they all belong to the one complete archaeological and historical heritage area known as The Llanymynech

Heritage Site. It would not be until 14th July 2006, some nineteen years later, that The Llanymynech Lime Works Heritage Area was scheduled under the Ancient Monuments and Archaeology Areas Act 1979. They were listed under the name of: the Lime Kilns, associated tramways, structures and other buildings at Llanymynech (The National Heritage of List for England, www.LlanymynechLimestonequarries/1021412, accessed 15/10/2013, Jones, 2004 and Ratcliffe, 2004).

All the modern track-way surfaces, protective fences, interpretation panels and direction indicators, benches and timber access stairs are excluded from the scheduling; yet the ground underneath them is included in the scheduling. The stone bridge which the A483 goes over is not included in the scheduling, but the Welsh and English tramways, which go under the bridge, are included. The wharves on the canal side form part of the boundary, as does the disused railway track of the scheduled area. Below is a map of the Llanymynech Limeworks Heritage Area (The National Heritage of List for England, www.LlanymynectLimestone quarries/1021412, accessed 15/10/2013.

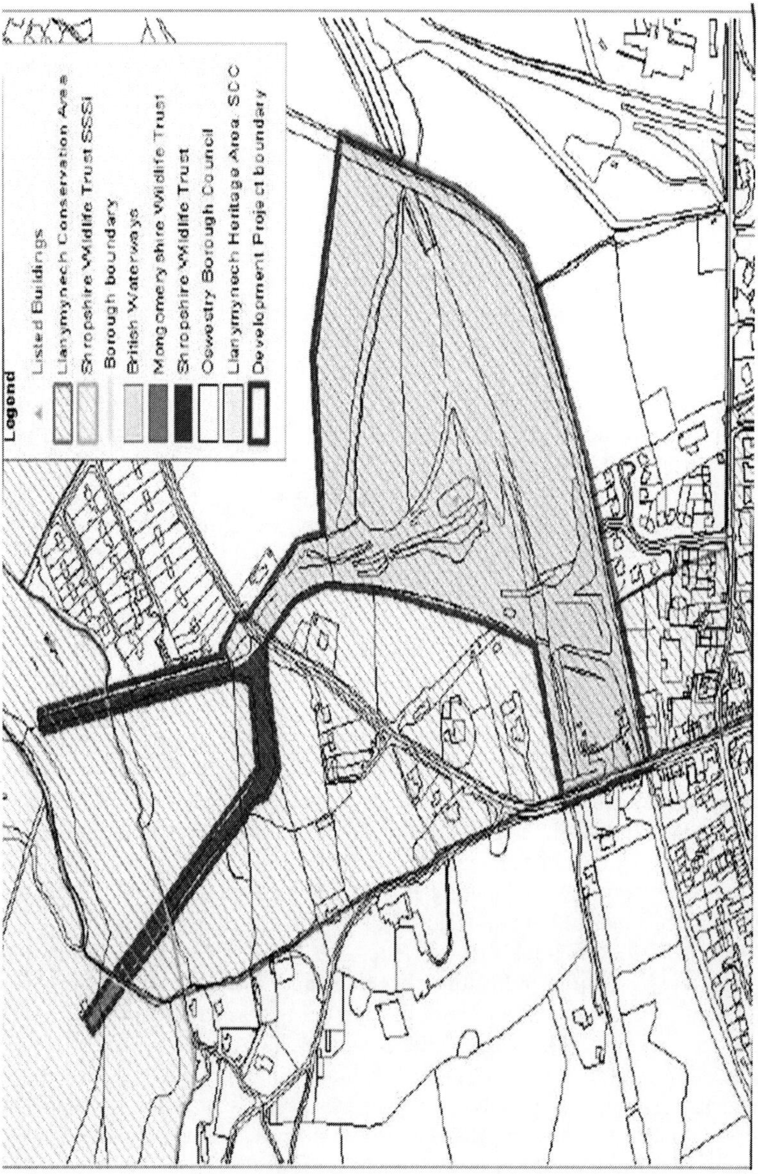

Figure 4.15: Llanymynech Limeworks Heritage Area Conservation Plan. (Source: Shropshire County Council accessed 08/11/2013).

4.10 Environmental Issues

Once industrial operation ceases at any quarry site, then nature reclaims the ground and this is the case with The Llanymynech Hill Quarries, as it has also been at The Much Wenlock, Lilleshall and Brown Clee Hill Quarries. Llanymynech and Llynclys Hills are Sites of Special Scientific Interest and are protected under Section 28 of the Wildlife and Countryside Act 1981. Virtually the entire hilltop of Llanymynech and its surrounding slopes, cliffs and quarries are designated as an (SSSI). Included in the (SSSI) are the extensive grasslands, bushes, forests and the natural rock faces, together with the abandoned Welsh and English Quarries and other past mining areas associated with copper and lead extraction (Jones, *et al*, 2012).

4.10.1 Llanymynech Rocks Wildlife Reserve

The former Welsh and English Quarries to the south and east of Llanymynech Hill were designated as a nature reserve more than thirty years ago, sometime around the late 1970s or early 1980s when they were named The Llanymynech Rocks Wildlife Reserve. The reserve is jointly managed by Shropshire and Montgomeryshire Wildlife Trusts. The reserve is of national and local importance and occupies the major part of the former limestone quarries. The importance of this site is in its limestone grassland flora and fauna which are described on information boards around the reserve. Following the closure of the limestone quarry nature has reclaimed the area; so for many decades extensive areas of Llanymynech Hill became overgrown essentially because of the removal of grazing sheep. This has been reversed following the reintroduction and controlled management of sheep to graze selected areas which will help clear the scrubland. Now that grazing has been undertaken, this will greatly enhance the visitors' perception and experience of The Llanymynech Rocks Wildlife Reserve. This measure has consequently also benefited the archaeological earthwork associated with the hillfort, making it easier to identify and appreciate the surrounding archaeology (Jones, *et al*, 2012).

The Llanymynech Wildlife Reserves contain very important flora and fauna showing the strength of nature when it is allowed to reclaim a site. When visitors enter the Llanymynech Wildlife Reserves they will first see the woodlands which include ash trees, wrapped in hanging creepers such as honeysuckle and wild clematis. As visitors follow the precipitous footpaths through the steep hilly reserve, they will cross a stile which leads into Wales and joins the Offa's Dyke footpath. From the top of the hill one can see for miles into the Shropshire plain, as far as the Breidden Hills in the Cotswolds, and Long Mountain with the River Vyrnwy winding below into Wales. The cliffs that were once the old quarry face now house nests of peregrine falcons, which can be seen flying about. There are also jackdaws and swifts in the summer (The Wildlife Trusts, www.shropshirewildlifetrust.org.uk, accessed 18/10/2013 and www.motwt.co.uk , accessed 18/10/2013). The Llanymynech limestone is very good for many plants and within this reserve are over three hundred and forty different species. The former quarry floor does not provide enough top-soil for grass to grow, however it does provide sufficient nutriment to give the less demanding plants a chance to grow such as primroses, cowslips and violets in spring.

On the hillside in June orchids flourish including The Greater Butterfly, Bee, Pyramidal and Common Spotted Orchids. In the final season of Autumn, Ladies Tresses and Autumn Gentian follows. On the violets and Bird's-Foot Trefoil plants Pearl-Bordered Fritillary caterpillars and Dingy Skipper caterpillars can be found eating. The quarry is still a suntrap as it was when a working quarry, however now it is not the quarrymen who are baking in the sun, but the basking lizards. Animals that like damp, secluded places, such as toads, frogs, slow-worms and newts, the latter being a protected species, are also found in these rock crevices; also perfect for hibernating animals. This describes only a small amount of the flora and fauna that can be found today (The Wildlife Trusts, www.shropshirewildlifetrust.org.uk, accessed 18/10/2013 and www.motwt.co.uk, accessed 18/10/2013).

Around the former The Llanymynech Lime Works site are woodlands and a canal which provide the ideal feeding ground for bats. The most significant feature of the lower part of the former lime works site are four species of resident British bats. Within, The Warren Hoffmann Kiln can be found Daubenton Bats who use the kiln as a maternity and roosting site. The Brown Long-Eared Bat has been seen flying close to vegetation, the woodland and around the kiln therefore it might be hibernating in the kiln. Other bats that have been seen in the area are the Common Pipistrelle Bat and the Whiskered Bat which roost mainly in old buildings. All bat species are protected by law under The Wildlife and Countryside Act, 1981. It is also an offence to do any work to a building that might disturb or harm the bats' roost; consent is needed from English Heritage before any work could be considered. Also bats can be disturbed by the noise of visiting groups entering into some of the buildings. Care should be taken at all times to advise visitors of the needs of the bats, before entering for example The Warren Hoffmann Kiln; also the numbers of visitors should be kept to a safe minimum (Ratcliffe, 2004 and Bat Conservation Trust, www.bat.org.uk, accessed 21/10/2013).

4.11 Social History

Research material found describes the social history of the people who lived and worked at The Llanymynech Lime Works. The last will and testament of Mrs Ann Dykes, widow of Llwynitdman in the Parish of Llanymynech, who died 16th December,1785, 67 years, leaves to her nephew, Mr Samuel Seley, all her rock tools, together with all the quarried coals and stones. Additionally left was the limestone prepared for lime burning together with limestone remaining on the quarry bank at her death. It is also interesting to note that when Mrs Ann Dykes died, she had been leasing lime kilns from Mr Richard Middleton of Chirk Castle.

She left instructions that her nephew, Mr Samuel Seley, should take possession and make use of these lime kilns (Joan Zorn, Per Comm, 04/04/2013), The second will relates to Mr John Edwards, a line-man, who died on 31st March, 1788, of 80 years, who also lived in the same parish. Mr John Edwards bequeathed to his grandson, Mr Edward Poole, all his rock tools of any sort, twenty pounds and his draught horse Mr John Edwards left to his son, Mr Henry Edwards, all the limestone that had been quarried and any that would be quarried in the future. He also left all the limestone 'lime coal' which remained unburnt at his quarry 'rockface' at the time of his death (Joan Zorn, Per Comm, 04/04/2013).

From my research, I was able to review the documentation of men recorded working in different jobs at Llanymynech Lime Works. Table 4.7 gives the names of the five men whose birth dates are recorded, but not their deaths. First is a Mr Thomas Griffiths, born 1802, who was resident in Llanymynech and Pant between the years 1841 to 1861. His job for the time he worked at Llanymynech Lime Works was described as rock labourer. Next is a Mr John Briggs, born 1853, whose record shows that in the year 1861 he was a scholar a person who had been schooled. Between the years 1871 to 1911 he lived at different places and changed positions every ten years. Mr John Owen, born 1816, again lived in Llanymynech or in Pant. Between the years 1851 to 1881 he worked as a rockman or rock labourer. A Mr John Lee, born 1834, is recorded in 1851 as living in Llanymynech and working as a rockman. The last person recorded is a Mr Edward Pryce, born 1799, who lived in Llanymynech and worked as a rock labourer and labourer in 1851, 1841 and in 1861 he working as a rock agent. Most of the jobs undertaken by these men were as rock labourers, apart from Mr John Briggs who was recorded has being a scholar in1861 and Mr Edward Pryce who worked his way up the ladder to the position of rock agent. Perhaps the work was hard to get at the time Mr John Briggs was recorded as a scholar and was it the only position he could find was at the lime works and stopped there.

One could possibly work out the ages of the five men by taking their last year of work from their date of birth. For example, Griffiths' last years of work was 1861. He was born 1802 so could have possibly lived to over 59 years. Similarly, Briggs' last year of work was 1911 and he was born 1853, so could have possibly lived till 58 years. Owen, whose last year of work was 1881, was born 1816, so again he could possibly have lived till approximately 65 years. Lee's last year of work was 1851; he was born 1834 so his age would have been approximately 17 years. The last man, Pryce, worked until 1861 and he was born in 1799, so he could have lived to approximately 62 years. This is the only documented evidence of these men's working life; however it does give an idea of their ages. Lee, 17 years old is the youngest person recorded working Llanymynech Lime Works at the time. What could have happened to him is unknown. He might have been involved in an accident or could simply have moved to another job. Despite the uncertainties, this is a unique insight to the people living and working in or near Llanymynech Limes Works. It is of interest that they lived such relatively long lives for the times, given that the conditions in which they will have worked will have been arduous (Source: www.Llanymynechlimestonequarries/4Ftons.tallyhouse, accessed 27/09/2013).

The following tables give the names of people who lived at Llanymynech or nearby and worked at The Llanymynech Lime Works:

Table 4.7: 1-5 Showing a number of men working at Llanymynech Lime Works

Name	Thomas Griffiths, born 1802 died?	
1861	Llanymynech, Residence	Rock Labourer.
1851	Llanymynech.	Rock Labourer.
1841	Pant.	Rock Labourer

Name	John Briggs, born 1853 died?	
1911	Gullet Cottage, Pant.	Rockman, Cambrian Railway.
1901	Mount Pleasant, Pant.	Rockman, Stone Quarry.
1891	Pant.	Lime Rock Labourer.
1881	Pant.	Lime Quarry Labourer.
1871	Llwynitdman.	Rockman.
1861	Upper Maesbrook.	Scholar.

Name	John Owen, born 1816 died?	
1881	Pant.	Rockman.
1871	Llanymynech.	Rockman.
1861	Llanymynech.	Rock Labourer.
1851	Llanymynech.	Rock Labourer.

Name	James Lee, born 1834 died?	
1851	Llanymynech.	Rock Labourer.

Name	Edward Pryce, born 1799 died?	
1861	Llwynitdman.	Rock Agent.
1851	Llanymynech.	Rock Labourer.
1841	Llanymynech.	Labourer.

(Source: www.Llanymynechlimestonequarries/4Ftons.tallyhouse, accessed 27/09/2013).

It is interesting note that Mrs Ann Dykes is recorded as working at The Llanymynech Lime Works in the middle to late 18th century; in addition she was leasing lime kilns from Mr Richard Middleton of Chirk Castle. Mrs Dykes was a widow, so it can be questioned as to whether she had inherited the rock tools and other items from her late husband and would have carried on working at the quarry after him. If there was one woman working at the quarry, it is quite possible that there were others. Another will, that of Mr John Edwards, shows that he left his grandson, Mr Edward Poole, all his rock tools and draught horse and twenty pounds; quite a lot of money in the late 18th century. Mr Edwards left his son, Mr Henry Edwards, all the limestone that had been quarried and any that would be quarried in the future. It seems that Mr John Edwards had a plan, by leaving his rock tools to his grandson and the limestone that needed to be quarried to his son, he would cause both men to have to form some sort of working relationship in order to benefit from their inheritances at The Llanymynech Limestone Quarries (Joan Zorn, Per Comm, 04/04/2013).

Table 7 provides a selected insight into some of the people, their employment and habitation, who worked at The Llanymynech Limestone Quarries. For instance, Mr Edward Pryce, born 1799, is shown starting his working life as a quarry labourer in 1841; in 1851 he was still working as a rock labourer in the quarries. However, by 1861Mr Edward Pryce had risen to the position of "Rock Agent", that is limestone quarry agent. This shows that a person could better himself in time (Source: www. Llanymynechlimestonequarries, accessed 27/09/2013).

4.12 Discussion

The Llanymynech Lime Works has been one of the most interesting sites to research, because of the amount of buildings still visible. After the closure of The Llanymynech Lime Works the site fell into disrepair and nature started to reclaim both the English and Welsh Quarries. The major problem with the Llanymynech Lime Works was that it was owned by seven different groups: The Shropshire County Council, The Shropshire and Montgomeryshire Wildlife Trusts, The Chirk Castle Estate, Earls of Bradford, British and Montgomeryshire Waterways. Mrs Harriet Devlin recognised the problem that was presented when too many individual groups own separate parts of a site. It is not always easy to co-ordinate organisations with slightly different interests. Mrs Harriet Devlin was fundamental in drawing together the differing groups to speak and work together for the conservation of the site. A Steering Group and Committee were initiated in order to form a management strategy for the whole of the former Llanymynech Lime Works Site, both the Welsh and English sides, so that foundations could be revealed, buildings and quarry sites preserved, and the regeneration of flora and fauna could begin. Currently, the Friends of Llanymynech have been vital in continuing to carry out essential work at the site such as, maintaining the footpaths, talking and telling the visitors of the enormous archaeological and historical background of the works (Harriet Devlin, Per Comm, 27/06/2013).

This major breakthrough has led to the conservation and preservation of The Llanymynech Lime Works, which is a unique site for both Wales's and England's industrial past; however it is an on-going project. Today the site offers educational opportunities: days out walking and exploring the site, canal journeys and railway trips on the Cambrian Heritage Railway. The Steering and Management Groups together with The Friends of The Llanymynech Lime Works are vital to the conservation, preservation and protection of the buildings and tramways from the quarries to the canal. The Friends also conduct walks around the former quarries, buildings and The Warren Hoffman Kiln of The Llanymynech Lime Works site.

Llanymynech, was I suggest was the major producer of limestone as agriculture fertiliser in Shropshire. There were other quarries producing limestone for agriculture fertiliser, such as those on Wenlock Edge, but not on the same scale. There limestone was used for other purpose as can be seen in the relearnt chapter.

4.13 Conclusion

Researching the former Llanymynech Lime Works has been very different from all the other quarries I have studied: Lilleshall, Wenlock Edge, Grinshill, Bayston Hill and Brown Clee Hill quarries. Llanymynech Hill and Lime Works was very different because of its position within the landscape, being partly sited in Wales and in England. In earliest times two different tribes inhabited the hill, one in Wales and the other in England. Llanymynech Hill has one of the largest Late Bronze Age, or Iron Age, hillforts to be found in England or Wales, thus this is another reason for its local and national importance. Over the centuries different groups of people have been quick to see the potential of the Llanymynech limestone, from the Romans into medieval times when it was attempted to mine silver in order to pay towards the ransom of King Richard I, 1193.

All of this has been fundamental in the archaeological and historical background of Llanymynech Hill, however the industrial period is of equal significance. The Llanymynech Lime Works played a vital part in the industrial period of the area and made it what it is today a heritage site. The uniqueness of the English and Welsh Quarries is that they were worked in such close proximity, while at the same time maintaining their separateness. The Welsh Quarries were owned by the Middleton's of Chirk Castle and the English Quarries by The Earls of Bradford, until Mr Thomas Savin leased both the Welsh and English Quarries at the same time. What makes this site so different from all the others is that there were separate Welsh and English Brake Drum Houses, inclined planes and canal wharves, all of which can still be seen today at The Llanymynech Lime Works Heritage Area which has been scheduled under the Ancient Monuments and Archaeology Areas Act 1979.

There is a wealth of records, relating to people who were part of the industry The Llanymynech Limes Works, not just the landowners, but the ordinary people who lived and worked in the quarries. Today in the 21st century local people are still very much associated with The Llanymynech Limes Works, in a different way now, telling of its industrial past in educational talks and guided walks.

CHAPTER 5

AGGREGATE QUARRIES: BAYSTON HILL (SHARPSTONE), BROWN CLEE HILL, CLEE HILLS

5.1 Introduction

This chapter will focus on two quarries; firstly Brown Clee Hill Quarry, completely exhausted by the early part of the 20th century and secondly Bayston Hill Quarry still working today in the 21st century. What both Bayston Hill and Brown Clee Hill Quarries have in common is the fact that they supplied aggregate stone for railways and for building and maintaining roads. Brown Clee Hill, at 540 metres above sea level, is the highest hill in Shropshire and is located near to Clee Village on the B4364 Bridgnorth to Ludlow Road. Bayston Hill Quarry, a working gritstone site, is located three miles South of Shrewsbury alongside the A49 to Ludlow and the mainline railway to the same destination.

5.2 The Geology of Bayston Hill, SO 493091 & Brown Clee Hill, Shropshire SO 593865

Igneous rocks form the basis of the crush-rock industry in Shropshire. The Precambrian and Dolerite volcanic rocks located on Bayston and Brown Clee Hills have been quarried principally for road stone. This rock produces high quality aggregates, suitable to be used as road stone as they meet the required abrasion and polish resistance. The most valuable aggregate deposits sought after in Britain are those with a rare high Polished Stone Value (PSV), such as the greywacke stone excavated from Bayston Hill which has a PSV of 65, thus making it ideal for road surfaces as it assists vehicles to brake more quickly. This is because the greywacke stone does not polish to a smooth slippery surface, but stays coarse and abrasive making it easier for tyres to grip (Toghill, 2006 and British

Geological Survey, Minerals Resource Information for Development Plans, 1998).

The approximate age of the sedimentary and volcanic rocks, the oldest rocks found in Shropshire, is 600 million years (Mya), and can be found on The Long Mynd, The Wrekin on Bayston Hill and Haughmond Hill (Toghill, 2006). At Bayston Hill, these rocks are formed from a series of mudstones and siltstones which overlie the Pre-Cambrian strata on the flanks of Bayston Hill on both sides of the quarry area and plunge towards the east, belonging to the Carboniferous period. The rocks on the south-eastern limits of Bayston Quarry are of the Keele Beds Division. The boundary between the Carboniferous Beds and the Pre-Cambrian strata beneath is defined by a significant unconformity and faulting does occur (Toghill, 2006 p69).

The bulk of the hill at Brown Clee is of Lower Old Red Sandstone capped by Carboniferous beds into which the Dolerite has intruded. The Lower Old Red Sandstone of Brown Clee Hill consists of 350 metres of green and brown sandstone; it is often recurring and bedded with only a few thin red marls and conglomerates. The outcrops of the Clee Hill Formation below the summit of Brown Clee Hill are of the Devonian Period, laid down some 420 to 360 million years ago (Toghill, 2006). The highest part of Brown Clee Hill was of the Carboniferous Period and comprises of igneous rock (dolerite or dhustone) which intruded some 360 to 290 million years ago into the Carboniferous Limestone and Millstone Grit. At Nordy Bank the limestone is some 4 metres thick; there are also two narrow bands of limestone of high purity of the St. Maughan's formation, known as the Abdon Limestone, an outcrop around Abdon Clee Hill (Toghill, 2006, Bryan, 2006 and Shropshire Archives, Notes on the Clee Hill Basalts, document no 524/1, accessed 22/12/2012). On Brown Clee Hill, these basaltic rocks, hard, black, often glassy, were formed from volcanic lava and are also known as Dhu stone, so named because 'Dhu' is the Celtic word for black" (Toghill, 2006 and Shropshire Archives, Notes on the Clee Hill Basalts, document no 524/1, accessed 22/12/2012). The basaltic lavas are sometimes comparatively thin, soft and partially weathered, so are therefore generally unsuitable to be quarried. However, the 60 metre Dolerite intrusion on Brown Clee Hill and predominantly on Titterstone Clee Hill has been quarried for road stone, and is still being quarried on Titterstone Clee Hill today. The greywacke stone quarried on Bayston Hill has a PSV of 65, whereas the dolerite quarried on Brown Clee Hill had a PSV of 52. This difference of 13 PSV makes the Brown Clee Hill stones very unsuitable as best surfacing material; yet it is otherwise very robust and durable, making it a hard-wearing quality stone, non-slippery in wet conditions (Toghill, 2006 and British Geological Survey, Minerals Resource Information for Development Plans, 1998). Below are the geological maps of Bayston Hill and Brown Clee Hill Quarries:

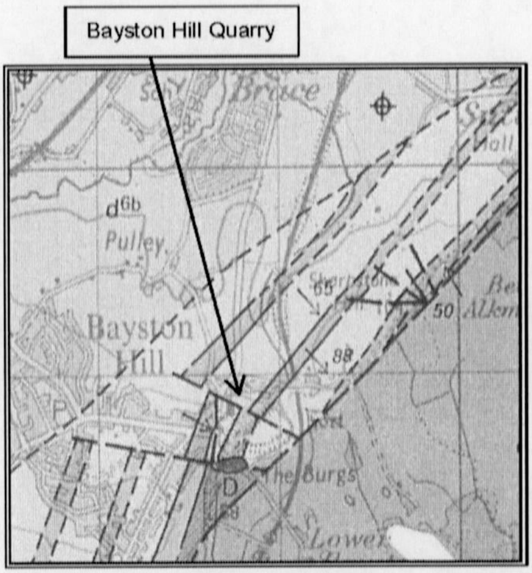

Figure 5.1: Bayston Hill Quarry Map of Mid Wales and Marches. (Source: British Geological Survey, accessed 14/11/1990).

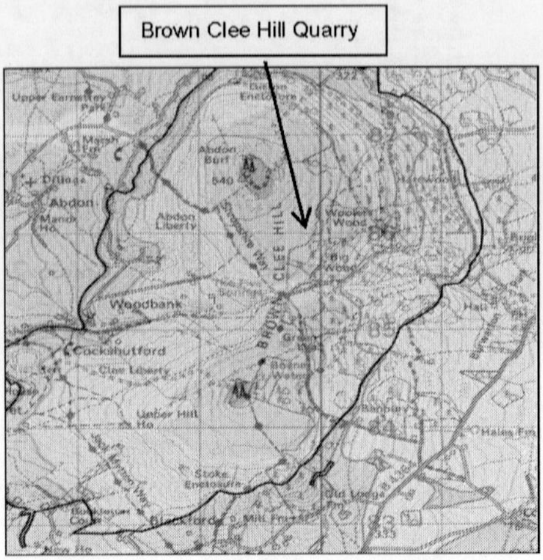

Figure 5.2: Geological Map of Brown Clee Hill, (Source: Geology Roam, Digimap.edina.ac.uk/geologyroam, accessed 19/12/2012).

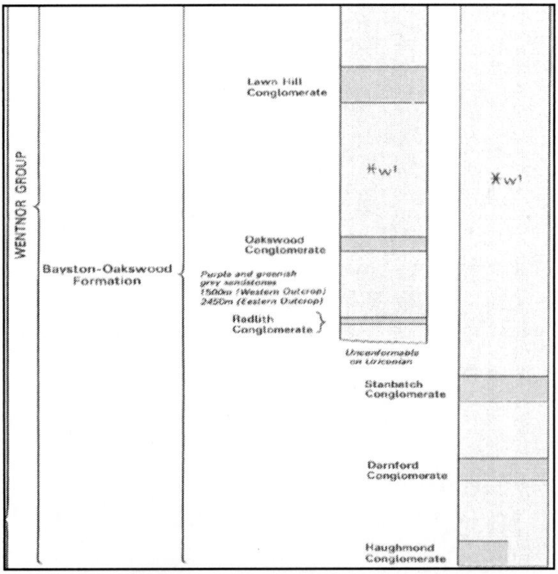

Figure 5.3: Map of Mid Wales and Marches: Index showing colours and symbols (Source: British Geological Survey, 1990).

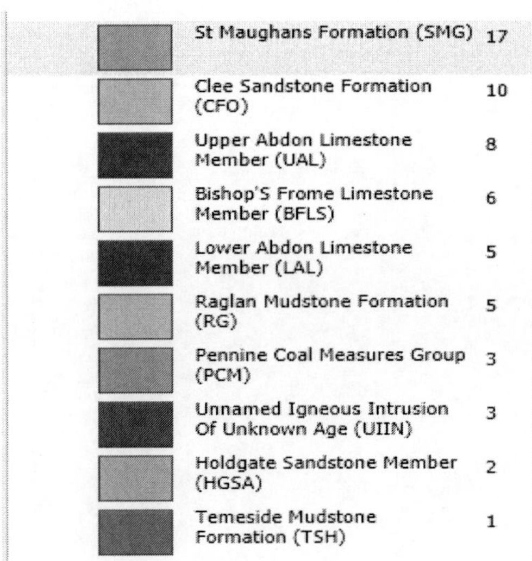

Figure 5.4: Geological Legend (Source: Geology Roam. Digimap.edina.ac.uk/geologyroam, accessed 19/12/2012).

5.3 Bayston Hill Greywacke and Brown Clee Hill Dolerite Qualities

Greywacke stone is generally referred to as an immature sandstone, consisting of poorly sorted, angular sand-to-gravel, sized particles. It is characterised by its grey colour and extreme hardness; it can have quartz veins and be high in iron, so giving it a reddish colour (Chisholm, 1911). The dolerite is a fine-grained coarse black rock or dark-greenish crystallised rock. The crystals in the dolerite can sometimes be up to 5 millimetres in length; dolerite is also rich in iron. The iron content in the greywacke and dolerite minerals makes them both very durable and resistant to weathering, thus ideal for road stone (British Geological Society, technical report WF/98/6, 1998, accessed 20/07/2011 and The Quarry, 1898).

The high proportion of iron and magnesium found in both greywacke and dolerite has significant consequences when working the quarries as it affects the physical properties of the stone. Both theses stones, because of their resistance to wear, make them ideal for specified use as rail ballast and research shows that The Bayston Hill Quarry was originally specifically set up to supply ballast for the railways (www.shropshire geology.org.uk, accessed 24/02/2011).

The greywacke stone quarried at Bayston Hill is also used locally as dimension stone, for building walls, as is the dolerite stone quarried from Brown Clee Hill. Dolerite stone from this quarry was also used as an aggregate for paving slabs and in the construction of concrete pre-fabricated buildings (British Geological Society, technical report WF/98/6, 1998, The Quarry, 1898, Shropshire Archive, Clee Hill Granite Company, document no, 632/1 and 632/3 accessed, 22/12/2012 and 30/11/2012).

5.4 History of Bayston and Brown Clee Hills

It is important, as far as possible, to outline the history of both Bayston Hill and Brown Clee Hill Quarries where many people were employed and which contributed to the economy of Shropshire for many years: in the case of Bayston Hill Quarry it still does so today.

On both Brown Clee Hill and Bayston Hill there is evidence for Prehistoric activity, from the Neolithic, Bronze Age and Iron Age periods. A 1998 archaeological evaluation of Bayston Hill discovered Neolithic and Bronze Ages ditches, a cemetery, middle Bronze Age barrows and also late Bronze Age or Early Iron Age field ditches. There was also an Iron Age hillfort; this is known as The Burgs with rectilinear crop mark enclosures found some metres north of this site. After excavations had taken place, it was discovered that the area had been occupied during the Iron Age and Early Romano-British periods (Bayston Hill, Archaeological Evaluation, Cotswold Archaeology Project 1998, accessed 02/09/2014; White and Wigley 2011). Crossing Sharpstone Hill from east to west is a Roman road which excavation suggests had Iron Age antecedents and overlay further Bronze Age evidence (Malim, *et al* 2011).

A Neolithic axe discovered on the side of Brown Clee Hill gives evidence of Neolithic activity in this area, indicating that there must have been hunting and possibly settlement at this site. Bronze Age Cairns are to be found on Brown Clee Hill; however some of these have been destroyed by the quarrying. There were three hillforts sited on Brown Clee Hill but the only one to survive today is at Nordy Bank. The other two were located at Brown and Clee Burf's and these were destroyed by quarrying (Shropshire Archives, document no, XLS7956, The Hill Forts of Shropshire, accessed 22/12/12 and Bryan 2006). None were properly excavated but if they are of similar date to the Titterstone Clee Hillfort then they will have been founded in the later Bronze Age continuing into the Iron Age (Barratt *et al.* 2007, 14-15). Abdon Burf, like Titterstone Clee hillfort comprises a simple bank and ditch, the bank being constructed from the loose dhustone rubble lying naturally on the hill surface. Further material will have been generated by the shallow excavation of the ditch. There is no evidence for stone extraction at this early date since the natural dhustone was too hard to extract using the technologies of the time.

The only Roman evidence discovered near Brown Clee Hill, in the Parish of Ditton Priors, is some Romano-British pottery dated 45 AD. Brown Clee Hill is mentioned in the Doomsday Book (1066) and tells how the lands passed from the Norman Earls to the Crown. The Crown then gave Brown Clee Hill with Ditton Prior to a Hugh de Perriers, who in turn willed it to Wenlock Priory, in 1176. However with the dissolution of Wenlock Priory in 1540, the lands were sold off by Henry VIII (Shropshire Archives, document no, OD 64 V. F., accessed 22/12/2012).

5.5 History and Owners of Bayston and Brown Clee Hills Quarries

It is important to establish as far as possible how quarrying was started and who owned both Brown Clee Hill and Bayston Quarries. It is important to establish, as far as possible, how quarrying started at the Brown Clee Hill and Bayston Quarries and who the owners of these two sites were. From research at Shropshire Archives and into Kelly's Historical Directories, information was found on the former owners of Bayston Hill Quarries during the 19th and 20th centuries. It started with a Mr William Fox in 1871 who was a farmer and quarry owner, the location of the quarry site has not been identified. In 1895, 1900 and 1905 Mr Peter Fox is recorded as being firstly a quarryman, who could have possibly been working for Mr William Fox. Then in 1900 and 1905 Mr Peter Fox is recorded as being the quarry owner at Bayston Hill. William Fox, owner of Bayston Hill Quarry, has a son Peter, or a nephew who works as a quarryman1895 learning the job from the ground up, from his father or uncle in time Peter takes over 1900 from his father or uncle and eventually takes over the ownership. During each year of 1909, 1913, 1917, 1922 up to 1944, Mr Charles Edward Locke is recorded as owning Bayston Hill. From the records he was the proprietor of Bayston Hill Quarry site for thirty-eight years. Had Mr Charles Edward Locke realised the potential of the Bayston Hill Quarry and its mineral greywacke, this could be the reason. From my research at Shropshire Archives, other places and from interviews, I could not discover any more information about Mr Charles Edward Locke and his connection with Bayston Hill Quarry. Tarmac Ltd's archives do have records of the Bayston Hill Quarries. Records of Mr Charles Edward Locke's connection with Bayston Hill Quarry may well be in Tarmac Ltd's archives, however, although I asked if I might see these archives my request was refused. In 1944 Messrs Harper and Thomas Ltd were the owners on Bayston Hill Quarries; they seem to have been the proprietors of the site until sometime in the 1950s. Tar Slag amalgamated in 1959 with Tarmac Ltd. For nine years between 1950 to 1966 Tar Slag were proprietors of Bayston Quarries. In the early 1970's Tarmac Ltd purchased Bayston Hill Quarry. At that time the owner of the mineral rights of the quarry was Mr John Adkins of Bomere Farm. In summary, it seems that possibly as early as 1909 the potential of Bayston Hill and its unique mineral was being comprehended. Today, major companies, such as Tarmac Ltd, are extracting large amounts of greywacke, and as said, this mineral is very rare and can only be found at certain sites in Britain (Historical Directories, www.historicaldirectories. org, accessed 03/03/2011, and Shropshire Archives, Kelly's Directory Shropshire, 1871, 1890, 1900 and 1920, accessed 03/03/2011).

Table 5.1: Bayston Hill Quarry Owners:

Fox, William.	Farmer and Stone Quarry Owner	1871, Bayston Hill, Sharpstone.
Fox, Peter.	Quarryman.	1895, Bayston Hill, Sharpstone.
Fox, Peter.	Quarry Owner.	1900, Bayston Hill, Sharpstone.
Fox, Peter.	Quarry Owner.	1905, Bayston Hill, Sharpstone.
Lock, Charles, Edward.	Proprietor, Tankerville Villa.	1909, 1913, 1917, 1922 and up to 1944 Bayston Hill Quarries.
Messrs Harper and Thomas Limited.	Quarry Owners.	1944, Bayston Hill Quarries.
Tar Slag (Tarmac).	Quarry Owners	1950s to 1966
Tarmac Limited.	Proprietors	Early 1972 Bayston Hill Quarries.

(Sources: Historical Directories, www.historicaldirectories.org, accessed 03/03/2011, Shropshire Archives, Kelly's Directory Shropshire, 1871, 1890, 1900 and 1920, accessed 03/03/2011 and Shropshire Archives , documents concerning the Township of Bayston Hill, accessed 24/02/2011).

Information drawn from the Historical Directories is given in this table which shows Bayston Hill quarries owners and farm owner. Similarly in Much Wenlock in 1841, Mr Richard Milner of Much Wenlock was a farmer and small quarry owner; likewise, in 1871, Mr Fix is also listed as a farmer and small quarry owners (Historical Directories, www.historical directories.org, accessed 03/03/2011 Williams, 1997).

Following my extensive research in the above references, it was possible only to find record of the owner's names, but no other facts were given for either Messrs Harper and Thomas Limited, or Tar Slag: I found no reference at all to these companies in the Shropshire Archives. So it was not possible to ascertain the precise date that they sold their quarry to Tar Slag; however from the 1950s until 1966, Tar Slag was operating the quarry when, for some reason, operations ceased. In 1959 Tar Slag amalgamated with Tarmac to form one company known as Tarmac Limited. In the early 1970s, Tarmac started to redevelop Bayston Hill Quarry, which they reopened in 1972 (Historical Directories, www.historicaldirectories.org, accessed 03/03/2011, and Shropshire Archives, Kelly's Directory Shropshire, 1871, 1890, 1900 and 1920, accessed 03/03/2011).

I placed an article in the May 2011 edition of The Villager, Bayston Hill's local magazine, asking any former or present quarry workers if they might be interested in talking about their connection to Bayston Hill Quarry, but received no reply. It was hoped that someone would have liked to have shared their knowledge and working life at Bayston Hill Quarry. A response to the advertisement placed in the local parish magazine, The Villager, provided contact with the former quarry manager of Bayston Hill Quarry from 1996 to 2000.

Brown Clee Hill offered natural resources such as timber to be coppiced, sandstone, coal, ironstone, limestone and dolerite, unlike the single resource, greywacke, quarried at Bayston Hill. From 1260 to 1263 Sir John de Halston was granted a licence by Walter de Clifford, Lord Corfham, to extract coal from Brown Clee Hill; the highest coal field in England. By some six centuries later in 1839, nearly all the coal had been extracted and the industry was in decline. It was replaced by the stone quarrying industry (Shropshire Archives, document no OD 64 V. F., accessed 22/12/2012, Shropshire Archives, Notes on the Clee Hill Basalts, document no 524/1, accessed 22/12/2012 and Hewitt, 1991).

Mr Thomas Roberts, an engineer returned from overseas, having settled in Ludlow, he focused his mind on developing the mineral resources on Clee Hills. So a group of like-minded people, John Mackay, William Clarke, Colonel's William Field and Patchett, joined Mr Thomas Roberts, as they appreciated the potential of the dolerite stone discovered on Brown Clee Hill which was ideal for use in the construction of roads. The people involved in this enterprise with Mr Thomas Roberts were already shareholders in other quarrying companies; The Clee Hill Dhustone Company, established in 1863, The Cornbrook and Knowbury Coal and Stone Company, c.1860, and The Titterstone Quarry Company established in 1881. The Clee Hill Granite Company was established in 1867 by Mr Thomas Roberts who managed the quarrying company on the behalf of the other shareholders. In the 1860s these farsighted men formed The Associated Stone Companies of Clee Hill; through which they were able to control the price of quarried stone. Mr Thomas Roberts died on the 25th August, 1887 and was succeeded by his son Mr Thomas Lee Roberts, Barrister-at-Law as proprietor of The Clee Hill Granite Company (The Quarry, 1889, Shropshire Archives, document no 5359/1, accessed 22/12/2012 and Bryan, 2006).

5.6 Other Minerals on Bayston and Brown Clee Hills

Before the middle of the 19th century there were no regular roads or waggon ways from the summit of the hills; the only way of getting the coppiced wood and quarried stone from the summit of the hill was by pack horse, on the backs of men and sometimes women.

Brown Clee Hill had other minerals at that time of equal importance, such as coal, iron ore and limestone. In medieval times ironstone and coal were mined on Brown Clee Hill; coal was mined by way of a "bell-pit" which is a narrow shaft opening out into a bell shape at its base. The bell-pits were very dangerous to work in because they were unstable and lacked ventilation. Today these bell-pits can still be seen on the Clee Hills (Barratt *et al.* 2007 and Abdon, A History of the County of Shropshire, vol 10, wwwbritish-history.ac.uk, accessed 08/04/2013).

5.6.1 Brown Clee Hill

Considerably more land on Brown Clee Hill has been in private ownership than that on Titterstone Clee Hill; The Burwarton Estate, in the ownership of Viscount Boyne possesses the eastern area and the western peripheries are owned by the Parish of Clee St. Margaret and numerous other land owners. Although the land on Brown Clee Hill was owned privately, some farmers' leases allowed them to work and burn lime; some form of limestone quarrying and lime burning had been taking place at this site since the late 1600s-1700s (Abdon, A History of the County of Shropshire, vol 10, wwwbritish-history.ac.uk, accessed 08/04/2013 and Bryan, 2006). From 1807-1810 there were lime works and kilns at Marsh Gate built inside the boundary on the edge of Abdon Common. The kilns were linked with three quarries, which can be seen on the 1890 map of Brown Clee Hill. The kilns associated with the limestone quarries were fuelled with coal from the Brown Clee Hill Coal mines. In 1841 limestone quarrying and lime burning began to be a sporadic occupation on and around Brown Clee Hill. Between the periods 1856-1870 a Mr Levi, a farmer, was recorded as being the last regular lime burner in the Brown Clee area. The 1861 census also records a Mr James Bowin of Ditton Priors, as being a lime man. However about the middle of the 19th century the limestone industry was in serious decline and was completely finished by the late 19th century. Finally all the lime kilns were in disuse by 1902 (Barratt *et al.* 2007, Abdon, A History of the County of Shropshire, vol 10, wwwbritish-history.ac.uk, accessed 08/04/2013 and Bryan, 2006).

5.6.2 Bayston Hill

The earliest recorded evidence of Bayston Hill Village is in the Domesday Book of 1086 (Shropshire Archives, documents concerning The Township of Bayston Hill, accessed 24/02/2011). In the 18th, 19th and 20th centuries, nearly all the land surrounding the village was used for agriculture and provided the majority of employment for the local people (Historical Directories. www.historicaldirectories.org accessed 03/03/2011). In 1844 Bayston Hill became part of the ecclesiastical parish with Lythwood Hall and part of Pulley, having formerly been part of the Parish of St. Julian and Condover; approximately three miles from Shrewsbury. A map of 1789 indicates 'stone crops', which could possibly suggest where the first quarry was worked. Shown on the same map is a turnpike road, which runs straight from Bayston Village to Shrewsbury; an ideal road for transporting quarried stone in the late 18th century, which is today known as the A49 and is still transporting quarried stone (Shropshire Archives, document no, 552/8/920, accessed, 17/03/2011). The map can be seen below:

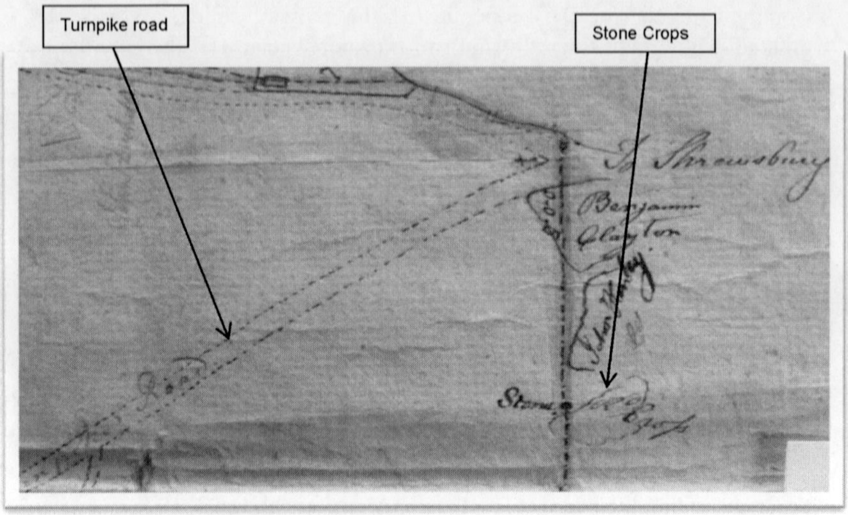

Figure 5.5: Map of Bayston Hill, dated 1789 Showing a Stone Crop. (Source: Shropshire Archives. accessed 11/03/2011).

In 1871, principal landowners in the Bayston Hill area were Mr William Henry Hornby and Mr Charles Kindersley. As shown in Table 1, between the periods 1871 to 1970 Bayston Hill Quarry was privately owned and operated. It is not clear how many people were employed on that site at that time. Tarmac Limited purchased Bayston Hill Quarry in the early 1970s, re-opening the quarry. The owner of the mineral rights at the quarry at that time was Mr John Adkins of Bomere Farm; at the same time Mr John Adkins was the tenant of Sutton Farm, owned by Tarmac Limited (Historical Directories, www.historicaldirectories.org, accessed 03/03/2011 and Shropshire County Council: Bayston Hill Parish Action Plan, September, 2006, accessed 17/03/2011).

Bayston Hill Common, once extended both sides of Lyth Hill and along Lythwood Road, on the eastern side of the village (Shropshire Archives, documents concerning The Township of Bayston Hill, accessed 24/02/2011). The remnants of the common today form the heart of the old and new village of Bayston Hill, which is divided by the A49. Bayston Hill Quarry is also linked to the A49 and near to the A5, which enables the processed stone to be transported across the country. It also has direct access to the railway which was used to transport its quarried stone until the 1950s when the railway sidings linking the quarry to the main railway-line closed. In 2007, Tarmac Limited made an application to Shropshire County Council, for the reopening of the railway sidings, which was granted. They reconsidered using rail transport because of the economics associated with road haulage. The new railway sidings linked Bayston Hill Quarry to the main Shrewsbury to Cardiff line, formerly the Newport to Gwent railway-line. On Lyth Hill in 1835 there was a rope works complete with a windmill. A church was built in 1843 on the eastern part of the common; this was to minister to the increasing population of quarry workers and railway navvies. Then four years later in 1847 a school was built for the education of the poor children of the parish (www.citizendia.org/Bayston _Hill, accessed 28/04/2011 and Shropshire County Council: Bayston Hill Parish Action Plan, September, 2006, accessed 17/03/2011). With the resumption of quarrying at Bayston Hill in 1972 special attention was given to the particular design and planning within the site in order to protect Bayston Hill Village and its local amenities. This was done by placing a primary crusher mounted on skids on the quarry floor, positioned as close to the quarry faces as possible; so cutting down on transport distances within the quarry (The Quarry Managers' Journal, 1974).

5.7 Quarrying on Bayston Hill and Brown Clee Hill in the 19th and 20th Centuries

An examination of maps from the 19th and 20th centuries will show how the quarry or quarries had expanded overtime until the present day at Bayston Hill and Brown Clee Hill, Shropshire. The 1880 and 1890 maps of Bayston Hill show five old quarries to the west and one to the east of the railway-lines. Over the decades, the three smaller and two larger quarries to the west seem to have remained the same size; as does the one eastern quarry situated near Sharpstone Lane. Both maps show buildings extant near the eastern quarry.

The 1900 and 1950 maps of Bayston Hill now only show three old quarries; two quarries to the west and one to the east of the railway-line. The eastern quarry, on the 1950s map, seems to have nearly doubled in size from all the previous maps. The 1950s map shows the area where a Stone Aged hammer was discovered in 1949. The 1970 map of Bayston Hill only shows one quarry, which is the site of the present quarry. Comparing the 1970 quarry with that on the 1950s map shows how the quarry has expanded in size over twenty years. The 1990 map of Bayston Hill indicates that the quarry increased by approximately one third over the subsequent twenty years period. Two sites on the 1990 map of this quarry, to the east and to the south are marked as 'quarry stores' (Digimap, www.digimap.edina.ac.uk/ancientroam, accessed 23/11/2011 and 30/11/2011). Bayston Hill maps can be seen below:

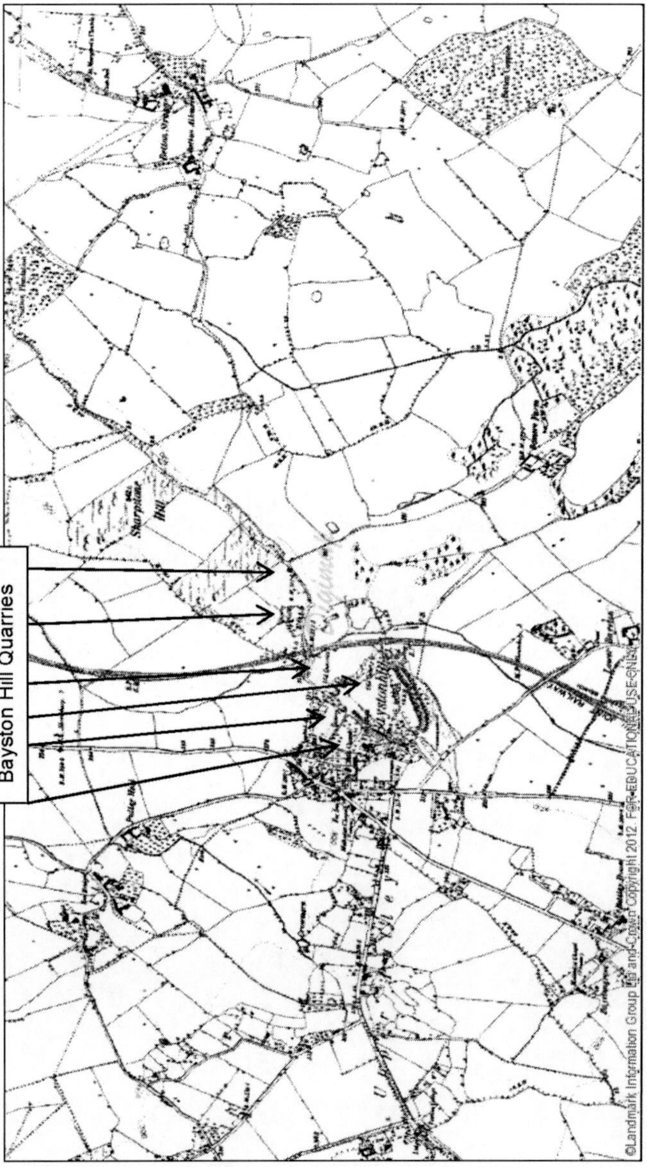

Figure 5.6: Bayston Hill 1880. Digimap. (Source: www.ac.uk/ancientroam, accessed 04/04/2014).

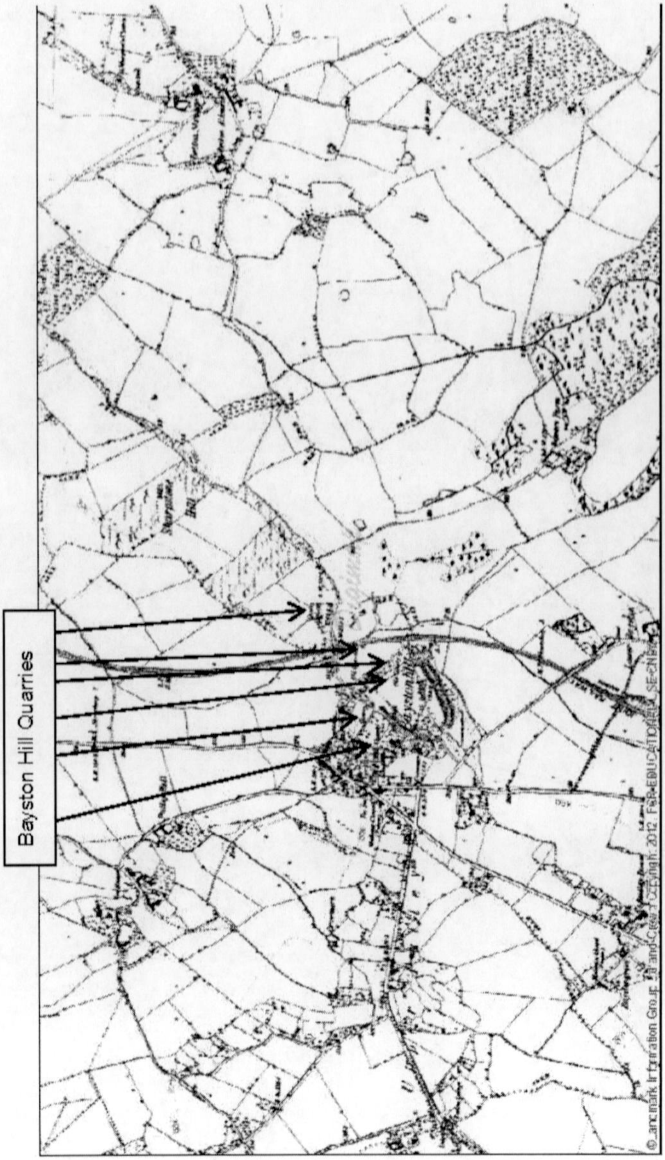

Figure 5.7: Bayston Hill 1890. (Source: Digimap, www.ac.uk/ancientroam, accessed04/042014).

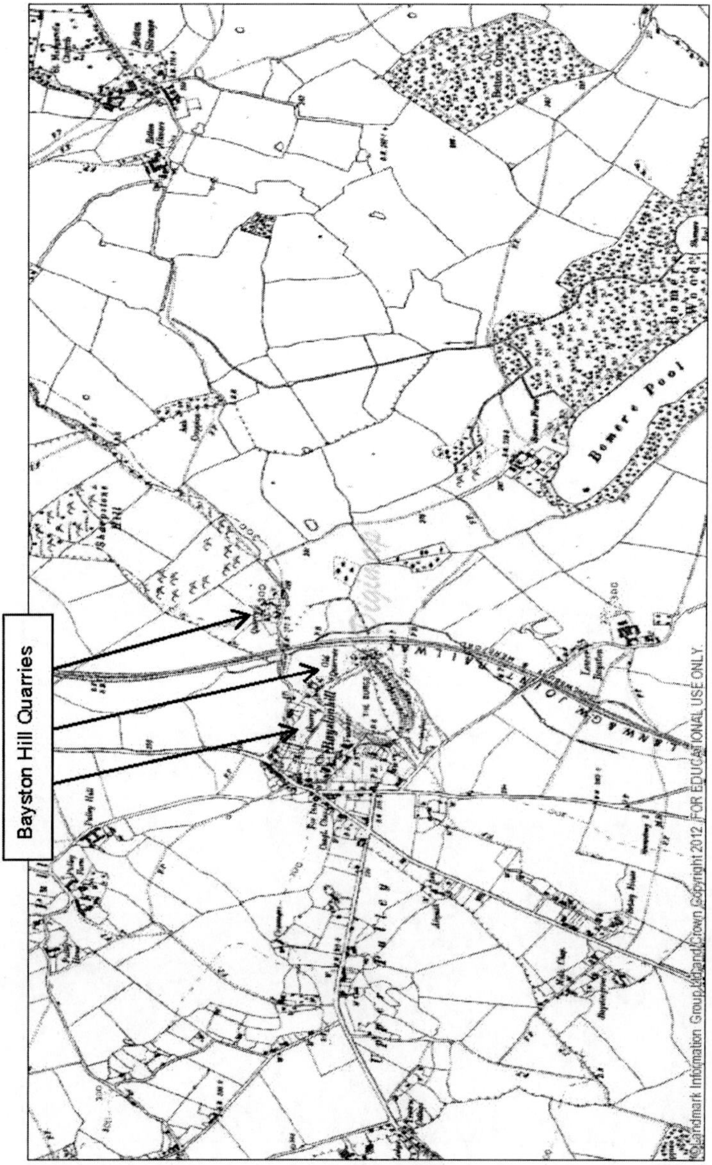

Figure 5.8: Bayston Hill 1900. (Source: Digimap, www.ac.uk/ancientroam, accessed 04/04/2014).

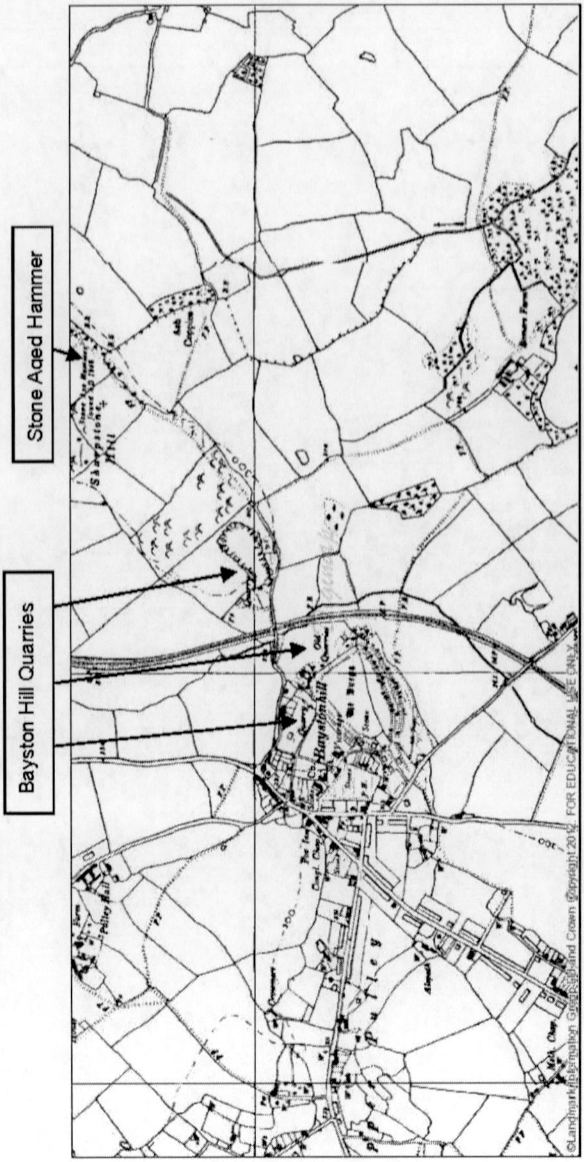

Figure 5.9: Bayston Hill 1950. Source: Digimap, www.ac.uk/ancientroam, accessed 04/04/2014).

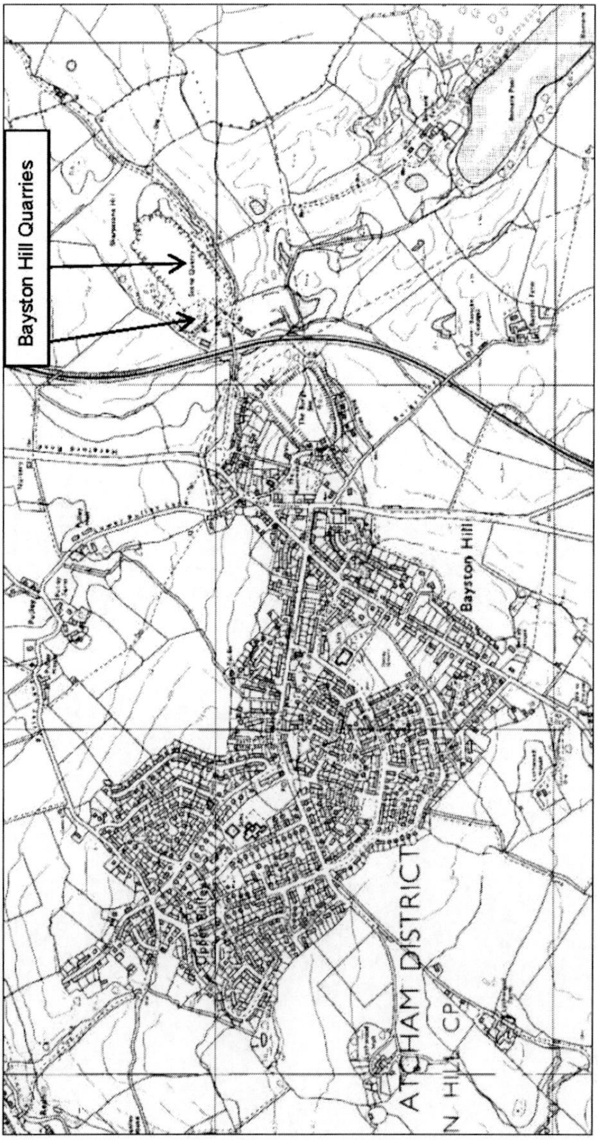

Figure 5.10: Bayston Hill 1970. (Source: Digimap, www.ac.uk/ancientroam, accessed 04/04/2014).

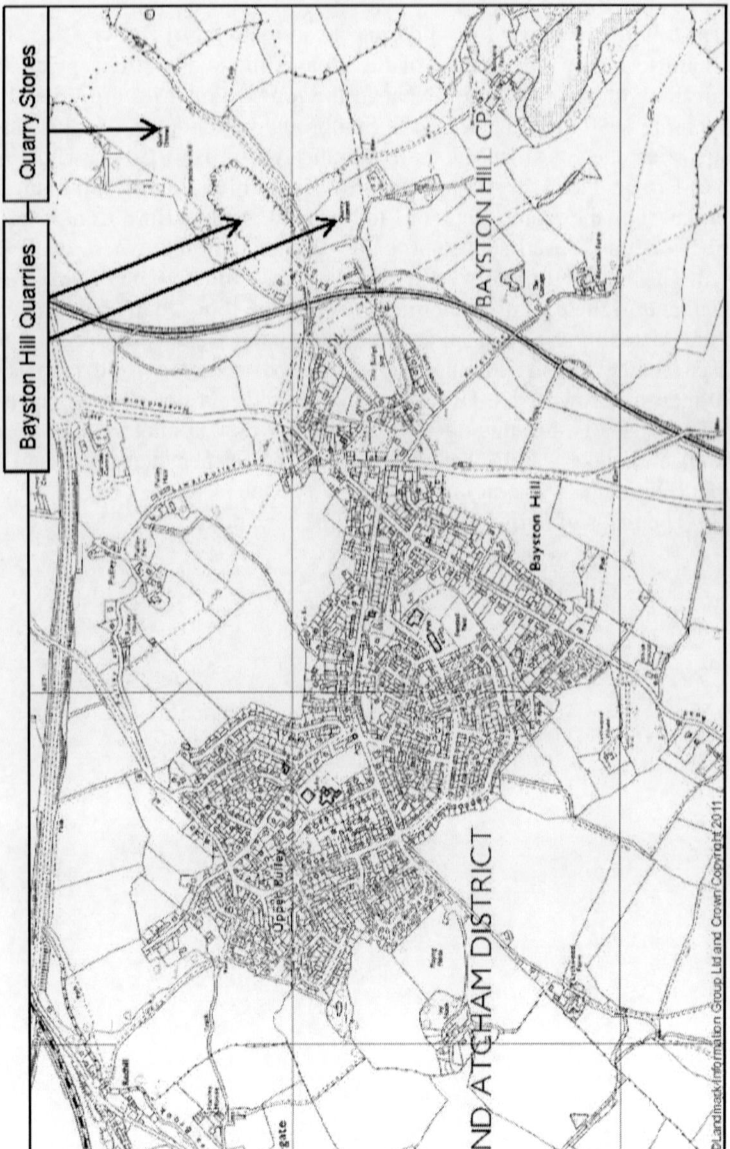

Figure 5.11: Bayston Hill 1990. (Source: Digimap, www.ac.uk/ancientroam, accessed 04/04/2014).

The 1890 map shows four old quarries to the west of Brown Clee Hill. It is hard to identify the extent of the area of the quarries, except for the one at the bottom of the map. Near to the top two quarries are limekilns. Also on the 1890 map, to the east of Stanbroughs Wood, are a further three old quarries and four disused shafts around Abdon Burf Camp. The 1900 map of Brown Clee Hill still shows these original four quarries, but not the limekilns. On the same map, on and around Abdon Burf Camp, are a number of old disused shafts, probably former bell-pits, where coal was mined. On the 1950 map, west of Brown Clee Hill, five old quarries are shown, and to the east of Stanbroughs Wood the four old quarries are still visible. Also still showing, on and around Abdon Burf Camp, are the old disused shafts, but the number has decreased from the 1900 map. The 1970 map of Brown Clee Hill now have no old quarries showing or any old disused shafts, but this does not mean that they are not visible, if one were to undertake a field walk of the area the former quarries would be visible (Digimap, www.digimap.edina.ac.uk/ancientroam, accessed 02/01/ 2013). The maps of Brown Clee Hill can be seen below:

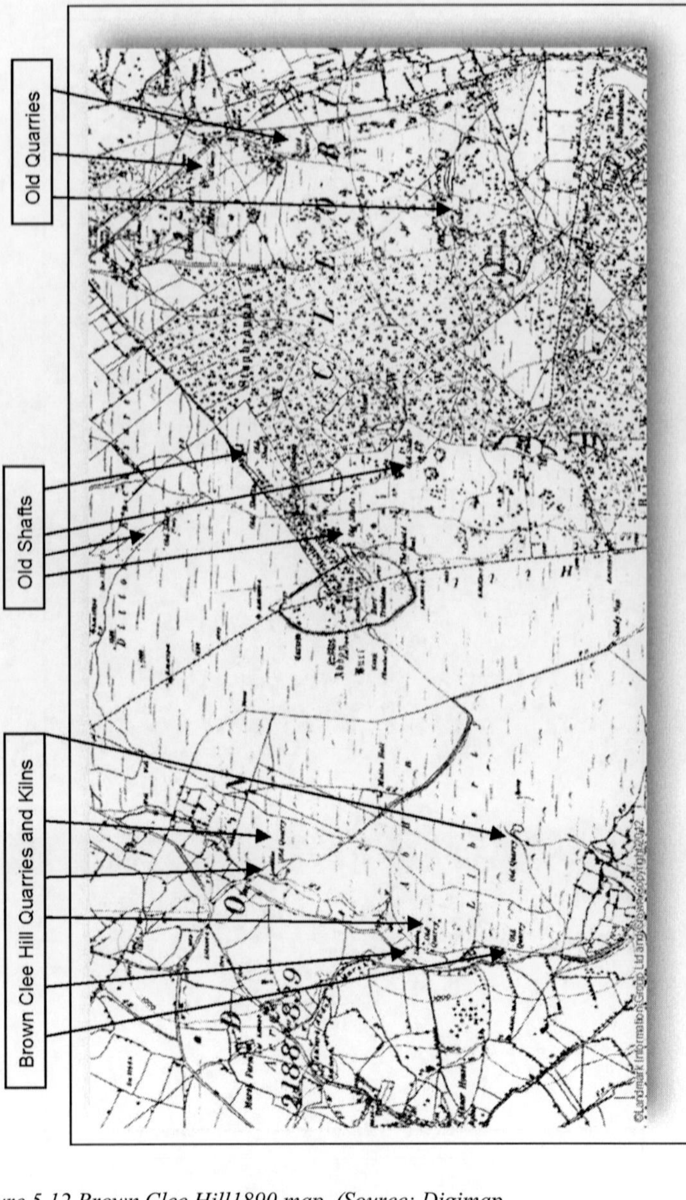

Figure 5.12 Brown Clee Hill1890 map. (Source: Digimap, www.ac.uk/ancientroam, accessed 04/04/2014).

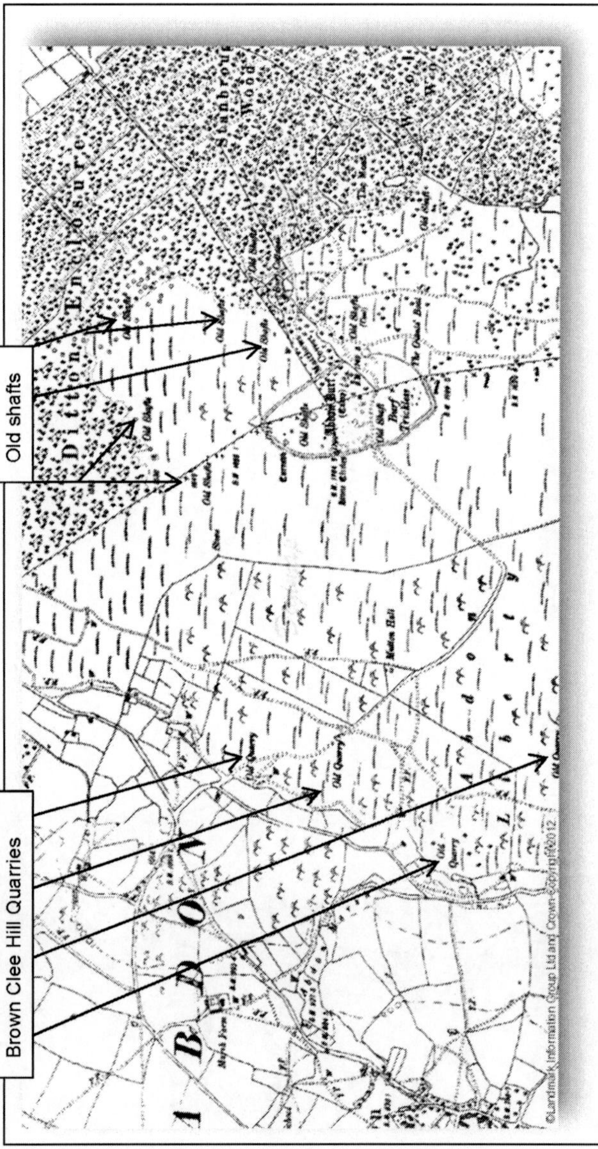

Figure 5.13: Brown Clee Hill 1900 map. (Source: Digimap, www.ac.uk/ancientroam, accessed 19/12/2012).

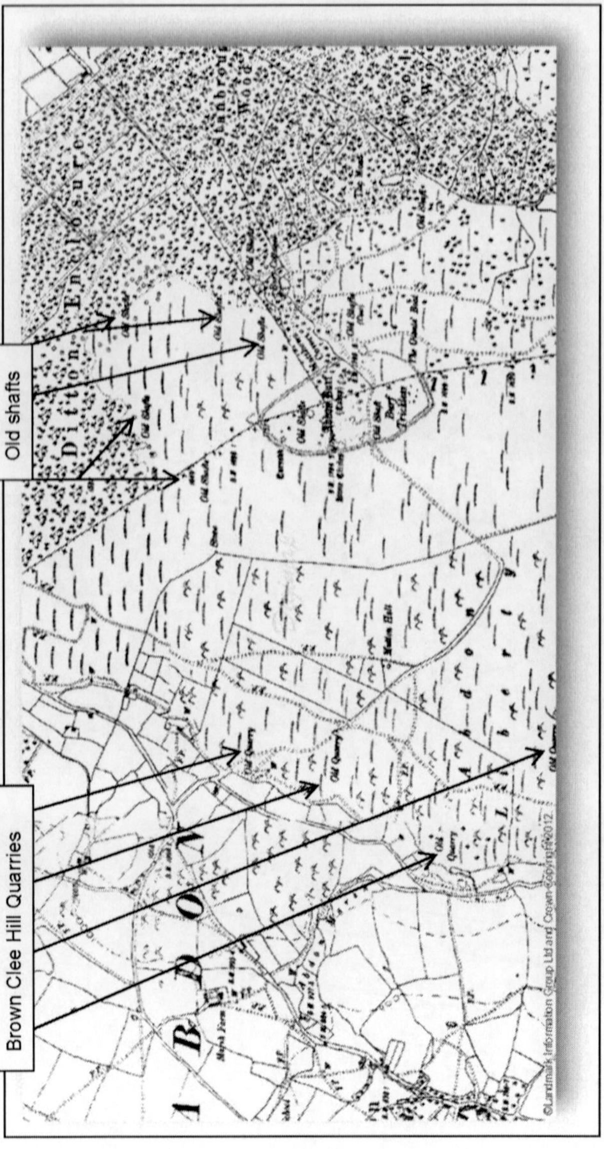

Figure 5.14: Brown Clee Hill 1950 map. (Source: Digimap, www.ac.uk/ancientroam, accessed 19/12/2013).

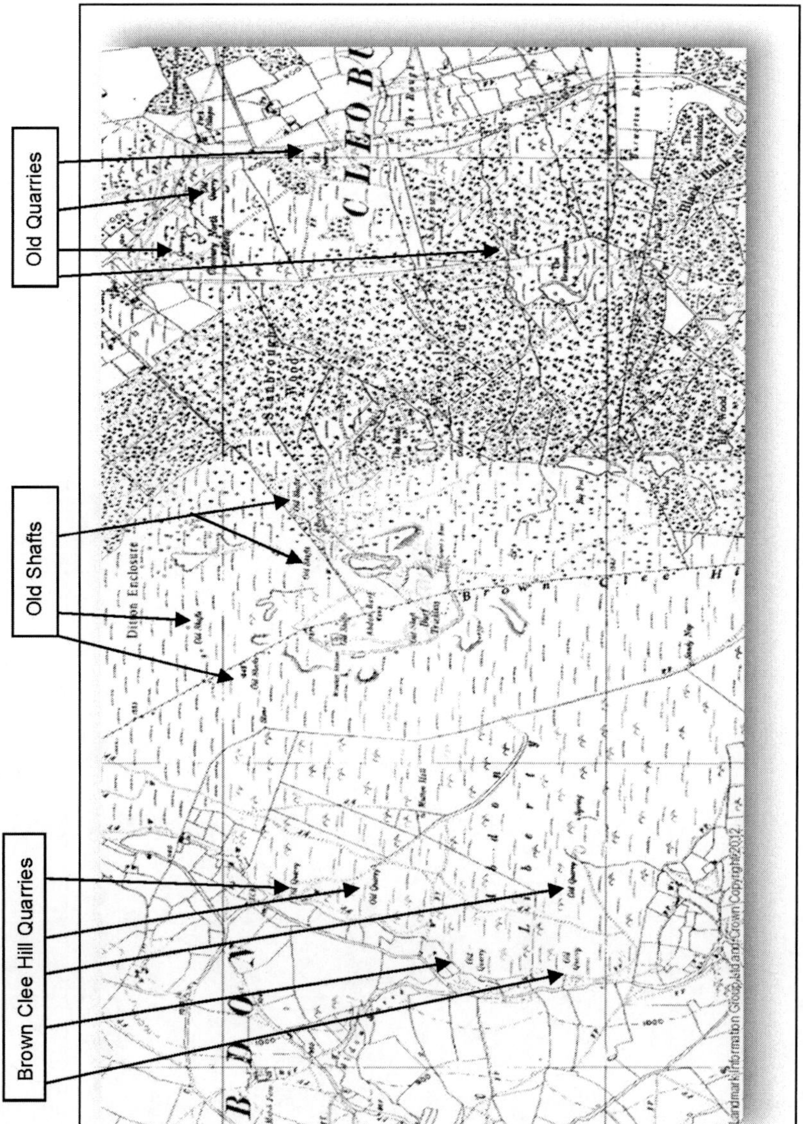

Figure 5.15: Brown Clee Hill 1970 map. (Source: Digimap, www.ac.uk/ancientroam, accessed 19/12/2013).

5.8 Brown Clee Hill Dolerite and Bayston Hill Greywacke Workings

5.8.1. Brown Clee Hill Dolerite Workings

5.8.1.1 Location

The difference in geological location between Brown Clee and Bayston Hill Quarries affected the prospects of Brown Clee Hill Quarries. The geological makeup of the Brown Clee Hill landscape affected any possibility of having a major railway system, although the Ditton Light Railway was constructed in the early 20th century; so railway transport was out of the question and even road transport was very difficult in the 19th and 20th centuries.

5.8.1.2 The Brown Clee Hill Quarry, Railways and Workforce

The men who worked at Brown Clee Quarries in the middle 19th and early 20th centuries started their day at 4am as they left their homes from the surrounding villages to walk to work at Brown Clee Hill Quarries, to start work at 6am; working a ten hour day which finished at 4pm. When the men arrived at the site offices, they would have another 45 minute walk up to the top of Brown Clee Hill; with a further hard climb to the quarry face. During the summer months when the daylight hours were longer, the quarry workers were expected to work double shifts. This was to build up a stock of quarried stone for the winter months when the weather was not so reliable; it would also allow the sett makers to work through these months. Setts were the cubed stones used to cobble streets, road and canal towpaths. The sett makers worked in a warm, well lighted shed (Shropshire Archives, document no OD 64 CF, accessed 22/12/2012 and Bryan, 2006).

A number of buildings were associated with Brown Clee Hill Quarrying Industry. The most important was the crushing and screening shed, followed by the blacksmith's forge, the carpenters' and fitters' shops and a wooden shed that housed one of the company's standard gauge engines. Also the "Whizz-bang Shed", which arose from the name given to the noisy locomotives used to pull waggons of quarried stone. When the primary sources of dolerite from the summit of Brown Clee Hill had been exhausted, the company looked for alternative sources. Investigations were carried out to locate new sources of dolerite near or around Brown Clee Hill, this was accomplished and called Crane Quarry, so named because of the large cranes, equipped with a skip to help move the excavated stone from the quarry (Shropshire Archives, document no OD 64 CF, accessed 22/12/2012 and Bryan, 2006). Below are pictures of the former Crane Quarry and other plant sheds:

Figure 5.16: Abdon Clee Hill Plant and quarry. (Source: Shropshire Archives, document no PH/A/16 accessed 30/11/2012).

Figure 5.17: Flood Quarry known as Crane Hole or Lake Kilowatt, on Abdon Burf, Abdon Clee Hill, Shropshire. (Source: www.geolocation.ws, accessed 12/04/213).

Figure 5.18: Frame of old quarry building on Abdon Burf, on Abdon Clee Hill, Shropshire (Source: www.geolocation.ws, accessed 12/04/2013).

Figure 5.19: Stone Crushing Plant on Abdon Clee Hill, Shropshire. (Source: www.geolocation.ws, accessed 12/04/2013).

Figure 5.20: Stone Crushing Plant, on Abdon Clee Hill, Shropshire. (Source: www.geolocation.ws, accessed 12/04/2013).

In the late 19th or early 20th centuries the Brown Clee Stone Quarry Company built a row of houses for their expanding workforce, situated on Hall Farm Road. Near to these houses were the stables for the quarry horses. The houses were just a step away from the quarry site, far from the days when the workers had to walk two hours just to get to the quarry offices. As any quarry worker's family occupying the new houses was obligated to the quarry company, so quarry workers would now have to think twice before taking industrial action because they would risk being given notice to quit their company homes. The consequences of their action would also affect the whole family. Casual workers were also employed by the quarry company, but they would have to find their own accommodation in the surrounding villages, or sleep rough. A local newspaper reported that a casual worker had been found frozen to death under a hedge (Shropshire Archives, document no OD 64 CF, accessed 22/12/2012, Bryan, 2006 and Price, 1976).

In the 1900s, Lord Boyne and other entrepreneurs were interested in further developing Brown Clee Hill Stone Quarry. They wanted to construct a light railway to transport the minerals, which would run from a junction with the Great Western Railway and Tenbury Bewdley Branch up to Ditton Priors, by way of a single track, where it would connect with an incline from Brown Clee Hill Quarry. The incline would be 243.84 metres in length from Brown Clee Hill Stone Quarry site to the Ditton Priors railway junction. The incline would help transport the quarried stone down the hill in 20 ton hoppers, on a counterbalance system. The company at this time employed some two hundred men (Bryan, 2006 and Price, 1976). Initially consultation on the route that the light railway would take was in 1900 when a Mr Everard Calthrop was appointed consultant engineer, but there were many difficulties until the 25th January 1907 when the long awaited construction started. On the 23rd March 1901, The Board of Trade confirmed the order to start construction of the 12.5 miles of the Ditton Priors Light Railway; however this was delayed until the summer of 1906. The delay in construction of the light railway was that all the tenders submitted were rejected. As a consequence, the original directors who had proposed the building of the Ditton Priors Light Railway seeing no return for their money, resigned from the board. Similarly people elected in their place also resigned after a short time (Shrewsbury Railway Heritage Trust, www.shrewsburyht.org.uj/history, accessed11/03/2013 and Bryan, 2006).

The original Board of Trade order made on the 23rd March 1901 to construct the Ditton Priors Light Railway line was planned to expire in 1906, so an application was made for an extension, which was granted. In January 1906 the board of directors received a tender from a Mr George Law of Kidderminster, to build the light railway within one year for £86,500 this would be split into £48,000 in cash paid to the above and £38,500 in shares. In May of 1906 the board of directors discussed Mr Law's tender, but by June 1906 no agreement could be made between both parties on the details of the contract, so subsequently the talks ended (Shrewsbury Railway Heritage Trust, www.shrewsburyht.org.uj/history, accessed 11/03/2013 and Bryan, 2006).

The Board of Trade were concerned about the length of time taken to get the project underway. In June 1906 a representative of Bott and Stennett Company attended a board meeting. Lord Boyne and his board of directors accepted the offer of Bott and Stennett Company, who had expressed their willingness to undertake the construction of the Light Railway line. Mr William Stennett was therefore appointed a director of The Cleobury Mortimer and Ditton Priors Light Railway during the period of its construction with a £100 salary per year (Shrewsbury Railway Heritage Trust, www.shrewsburyht.org.uj/history, accessed 11/03/2013 and Bryan, 2006).

Before the main work of laying the railway-lines could be undertaken it was necessary to build a bridge, divert a small brook and construct a new culvert. The major job was excavating and preparing the land prior to railway-lines being laid. A further application was made to The Board of Trade for an extension, which was granted until the year ending 1908. Finally, after eight years The Cleobury Mortimer and Ditton Priors Light

Railway was completed and opened on the 20th November 1908; first for goods traffic and later that year for passengers (Shrewsbury Railway Heritage Trust, www.shrewsburyht.org.uj/history, accessed 11/03/2013 and Bryan, 2006).

The life of The Cleobury Mortimer and Ditton Priors Light Railway was quite short, being absorbed by The Great Western Railway Company in 1921; some 17 years later in 1938 saw the last passenger train, the line finally closed in 1939. The Royal Navy, which had a Royal Navy Armament Depot near Ditton Priors Village, used the light railway line until it closed for good in 1965 (Shrewsbury Railway Heritage Trust, www.shrewsburyht.org.uj/history, accessed 11/03/2013 and Bryan, 2006).

5.8.1.3 The Workforce

The skilled workmen who worked at the quarry face excavated the stone, which the labourers would then load into waggons; these were then pulled along rail-tracks by individual horses to the crushing shed. The processed stone was transferred in waggons by small steam locomotives to the drum-head; this controlled the descent down by incline to the Ditton Priors Light Railway. There were a variety of jobs at the Brown Clee Hill Stone Quarry; quarrymen at the face, plant operators, train drivers, general labourers loading the waggons, farriers caring for the horses, blacksmith for the maintenance of waggons and rail-tracks. By the end of the First World War, the original traction power of horses had been replaced by trains to transport the processed stone from Brown Clee Hill Stone Quarry. After 1918 horses were still used, but in a subsidiary role at the quarry face (Shropshire Archives, document no OD 64 CF, accessed 22/12/2012, Bryan, 2006, pp., 94-97).

In the early 20th century the Brown Clee Hill Stone Quarry Company looked to divert its resources into another industry. This was achieved from 1914 to 1941, when the company started a ferrocement concrete works near Ditton Priors Railway Station. The process of Ferrocement or (Ferro), reinforced concrete, was developed by Hamish Cross. This is a mixture of cement and sand which is then applied over layers of expanding steel mesh with closely spaced small diameter steel rods. The new works employed 200 men whose job was to cast panels and posts for sections in buildings. In 1921, The Ditton Priors concrete works were very prosperous and a Mr Harold Bradley was earning £1.4.0 per week. These new works benefited the local area's economy. The finished parts were stored in a dead end siding at Ditton Priors Railway beside the wharf, waiting to be loaded by crane onto waggons (Shropshire Archives, document no OD 64

CF, accessed 22/12/2012). Clee Hills quarried stone has been used in many local buildings such as, The Clee Hill Village Hall, the church, The Golden Cross, Royal Oak, Angel Pub, squatter's cottages and some farmhouses. The dolerite, known locally as dhustone, is very hard to work and because of this the finished stone is of irregular shape, so producing irregular bonding (Shropshire Archives, document no OD 64 CF, accessed 22/12/2012).

An early 1936 document provides evidence indicating that 'the writing was on the wall' for Brown Clee Stone Quarries (Shropshire Archives, document no OD 64 CF, accessed 22/12/2012). At that time The Cleobury Mortimer and Ditton Priors Light Railway were only two trains per day running from the junction with the quarry incline. In 1936 or 1937 evidence suggests that quarrying ceased altogether on Brown Clee; which could mean that different quarries were worked out by 1936 and then the last quarry known as Crane Hole or Lake Kilowatt finished in 1937. After this a company called Simms-Lewis dismantled the equipment left on Brown Clee Hill by the former quarry company. Some of the quarrymen from the Brown Clee Hill Stone Quarry moved to the Cockshutford Quarries on the other side of the Brown Clee Hill; though the dolerite was not of the same quality and so the working did not last (Shropshire Archives, document no OD 64 CF, accessed 22/12/2012, and Bryan, 2006).

Prior to the First World War a small tarmac plant was established on the Brown Clee Hill. This was successfully developed using fine granite mixed with tar and was named Asphalt. The granulator machine, powered by a Sentinel engine, received the stone that had already been crushed to make fine chippings. In 1933, the Brown Clee Hill Stone Company built a larger tarmac plant at Ditton Priors, thus replacing the small one on the hill, making the process more economical and profitable for the company (Shropshire Archives, document no OD 64 CF, accessed 22/12/2012 and Bryan, 2006).

5.8.1.4 The Highway Act of 1862

The Highway Act, 1862 which provided guidance for surveyors, also advised that the dark basalt "dolerite" of Clee Hills, Shropshire, was near-perfect stone. This was because of its high portion of hornblende, which consisted of silicates, magnesium, iron, feldspar and quartz, the latter two being the main components. The government surveyors suggested that it can be a false economy to use softer stone, simply because it may be obtained nearby and cheaper. Road construction and maintenance in the

19th century was in its early stages, so it was not surprising that surveyors were not aware of the need for hard stone with hard-wearing qualities. So they used inferior stone that was easily and locally sourced. The argument that had been put forward was that the higher quality stone such as, greywackes and dolerite, was difficult to obtain locally if not impossible, so that the combined price of higher quality stone and its transport made it unattainable at an equal or lower price to the softer stone (Shropshire Archives, The Clee Hill Granite Company, document no 632/1, accessed 22/12/2012).

5.8.1.5 Testimonials of the Quality of Brown Clee Stone and its usage

Both Brown Clee Hill Dolerite and Bayston Hill Greywacke are among the hardest minerals which are shown in their PSV of 52 and 65 respectively. The dolerite stone of Brown Clee Hill was used for tramways setts and road surfaces across Britain; however the company also supplied rough or broken stone, for other purposes such as chipping for parks, drives and kerbing.

Summarised here are a few testimonies received by the Brown Clee Hill Stone Company from clients satisfied as to the quality of the stone. A Mr Ralph Hales for Lawton, Burslem and Newcastle-under-Lyne, Turnpike Roads, Surveyor's Department, Cobridge, Staffordshire Potteries throughout the 1880s had used Clee Hill Stone for the purpose of macadamising roads and paving slabs [i.e. setts] in Staffordshire. The stone surpassed in hardness any other stone they had ever used and they were very satisfied (The Shropshire Archives, The Clee Hill Granite Company, document no 632/1 accessed 22/12/2012). Mowlem, Freeman and Burt's Company of Westminster, London, declared in a letter of 1885, that the Clee Hill dolerite was an exceptional stone for macadamising roads, better than any other stone they had used. They stated that they were now supplying a considerable quantity of Clee Hill Stone for the paths and roads of London, where its hardness and durability had been noted (The Shropshire Archives, The Clee Hill Granite Company, document no 632/1 accessed 22/12/2012). Mr K. F. Clarke of The Highway District Surveyor's Office, Ludlow, commentated that in his twenty-five years he had found no other stone to be the equal of Clee Hill Stone. In addition, he stated that once the road had been constructed with Clee Hill Stone, maintenance and the employment of labourers had been considerably reduced (The Shropshire Archives, The Clee Hill Granite Company, document no 632/1 accessed 22/12/2012).

In the years 1875 to 1879, the Bishop's Castle Highway Board had used approximately 60 tons of local stone per year on road construction. However, from 1883 to 1886 this trend altered and the board started to use more Clee Hill Stone. One can also see that spending was reduced during that seven year period; subtracting the 1886 total of £1071:15:0 from the 1880 total of £608:11:04 leaves a reduction in expenditure of £463:03:08. The yearly cost of manual labour was reduced by £63:08:02 over the same seven-year period, recording a saving of £443:17:02. In 1886 the Surveyor, Mr A. Hamar, to the Bishop's Castle Highway Board, stated that everywhere Clee Hill Stone had been used in the Bishop's Castle District, it had given excellent satisfaction for macadamising roads and had saved the board money.

The chart in table 5.2 shows the stone used by The Bishop's Castle District Highway Board. The legend is blue for Clee Hill, red for local and green for the total combined stone used. One can see from the chart that in 1880, the blue column registers approximately 700 tonnes, with the red 2,600 tonnes approximately, and the total in the green column combines to 3,300 tonnes. There is a difference of approximately 1,900 tonnes between the blue and red columns. The column for the year 1881 shows that the blue column has increased to approximately 800 tonnes, whereas the red column has decreased to approximately1,450 tonnes, and the combined total in the green column is approximately 2,600 tonnes used. This shows a difference of 850 tonnes between the blue and red, even the green total of stone used has fallen by 700 tonnes. In the year 1882 the blue column seems to be approximately 800 tonnes, the same as the previous year. The red column has decreased again too approximately to 1,050 tonnes, where the combined total in the green column is approximately 1,850 tonnes. From a usage of 2,600 tonnes in 1880, 1,050 tonnes in 1880, in1882, the red column has again decreased by 250 tonnes, showing a steady decline in the local stone being used. In the year 1883, usage of stone recorded in the blue column, which is the Clee Hill Stone, has now over-taken the local stone for the first time. The blue column for 1883 now shows a slight increase by approximately 825 tonnes, whereas the red column records usage of approximately 600 tonnes. The blue column now shows an increase of approximately 225 tonnes of Clee Hill stone being used. The green column for the year 1883 now shows an approximately combined total of 1,495 tonnes. The blue column for the year 1884 seems to have stabilised to 825 tonnes, the same as the year 1883. However the red column for the year 1884 continues to decline to approximately 475 tonnes. The blue column has again increased by approximately 350 tonnes, where the combined total in the green column is approximately 1,250 tonnes.

The Blue column for the year 1885 shows a little increase of approximately 830 tonnes, where the red column has continued to decline to approximately 200 tonnes. The blue column has increased by 630 tonnes of Clee Hill stone. The green total column is now 1,100 tonnes of stone used. The last year, 1886, shows that the blue column has decreased to approximately 700 tonnes, a fall from the previous year of approximately 130 tonnes. This is the first time there has been a decrease in the stone from Clee Hill over the period 1880 to1885. However the red column continues to decrease to approximately 175 tonnes, a difference of approximately 550 tonnes from the blue column. The green column continues to decrease to approximately 1,495 tonnes. This chart shows in the red column that local stone used by The Bishop's Castle District Highway Board had declined rapidly from 1880 when the local stone used was approximately 2,600 tonnes, to approximately 175 tonnes in 1886, a decline of approximately 2,425 over the period of six years. Whereas the Clee Hill stone in the blue column increased after the year 1880, in the years from 1881 to 1885 usage was either the same as in the previous year, or increased, as in the year 1886. In the years 1880 to 1882 local stone was being used more as can be seen in the red column. However, as usage of the local stone decreased so by comparison there was an increase in the uptake of Clee Hill stone. During the years 1883 to 1886, Clee Hill stone was being used in preference to the local stone. This shows that The Bishop's Castle District Highway Board preferred Clee Hill stone, which could have been of a better quality to the local stone. However the combined total, shown in the green column, shows that over the period of seven years the combined use of stone from both local and Clee Hill had declined. This is shown in the chart's green column when the two totals from 1880 and 1886 are taken away from each other. The 1880 total in the green column is approximately 3,300 tonnes, and approximately 1,495 in the year 1886, a decrease of approximately 1,805 tonnes. I could not find any archive information for the reason why there had been a decline in the stone used in highway construction or maintenance. One possibility is that the Bishop's Castle District Highway Board had found another way of highway building and maintenance (The Shropshire Archives, The Clee Hill Granite Company, document no 632/1 accessed 22/12/2012). Shown below is table 5.2:

Table 5.2: Stone used by Bishop's Castle District Highway Board

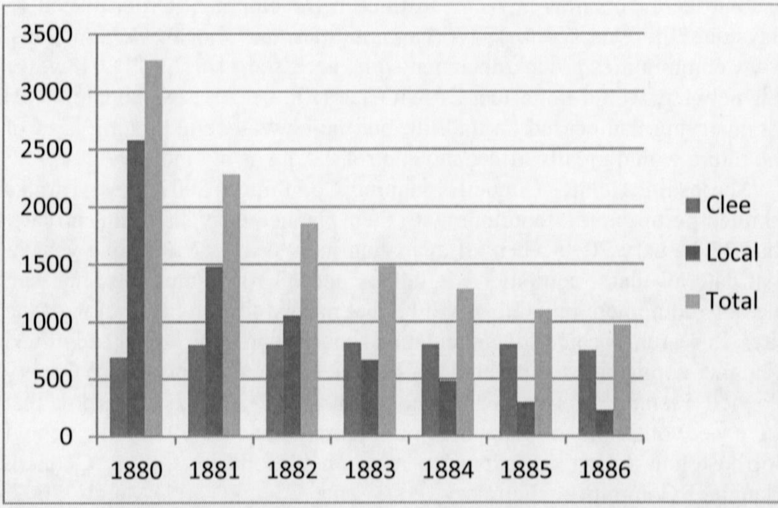

(Source: Shropshire Archives, The Clee Hill Granite Company, document no 632/1, accessed 22/12/2012).

5.9.2. Bayston Hill Greywacke Workings

5.8.2.1 Greywacke

On Bayston Hill from 1789 to the present day, there has been some form of quarrying, firstly rudimentary and now highly technology based; permission has been granted for quarrying to continue until 1st January 2060. The geological location of the minerals defines the position of each individual quarry or mine. In the case of Bayston Hill Quarry's location, it is not only sited immediately over the mineral quarried, but it is also connected by railway and by road to the A49 which in turn leads on to a network of major roads and motorways. Bayston Hill Quarry is rated one of the best stone quarries in England as its product, greywacke, is a very sought-after commodity, of which there are limited resources (Shropshire County Council, Planning Committee, February 2001 and 7th and 20th March 2007: MS2003/0529/SY accessed 04/04/2011 and British Geological Survey: The Economic Importance of Minerals to the UK, www.minerals UKcom, accessed 06/04/2011).

The Town and Country Act, 1947 provided the first control over the development and usage of land in the country. Before this Act of

Parliament, the use of land for quarrying and development was uncontrolled. This new law saw an Interim Development Order placed on Bayston Hill Quarry in 1947 (Communities and Local Governments, www.communities.gov.uk/documents/pla, accessed 11/07/2011). However this new law would not affect Brown Clee Hill, because by the late 1930s all quarrying had ceased on the hill; but the new law and coming laws of the future would greatly affect and control the quarrying industry.

Shropshire County Council, Planning Committee, 2007, gave Tarmac Limited permission to continue with their planned extension until no later than 1st January 2060 when all quarrying must cease. Within one year of that date all plant, equipment, buildings and all roads must be removed; the only equipment remaining will be that needed for the restoration of the site. Research located documentation concerning four other approved planning applications submitted by Tarmac Limited to Shropshire County Council, Planning Committee for the periods 1992-2007. It is possible that there were other planning applications summited before 1992 which could not be found during my extensive research (Shropshire County Council, Planning Committee, February 2001 and 7th and 20th March 2007: MS2003/0529/SY and MS2005/1165/SY accessed 04/04/2011).

5.8.2.2 Environmental Considerations

In all Tarmac Limited applications for extension of their quarrying rights at Bayston Hill major environmental considerations were found. Shropshire County Council always provides guidelines before any permission is granted for a quarry extension and these have to be instigated by the quarry owners:

- Environmental Considerations
- Effects on Local Residents and Amenities
- Blasting, Noise and Dust Effects
- Operation Hours
- Road and Rail Transport
- Landscape Impact
- Ecological Considerations including effects on Sensitive Sites and Animal Species
- Public and Access Rights of Way
- The Effects on Agricultural Land
- Archaeological and Cultural Heritage
- Restoration of Quarry Site and Aftercare

5.8.2.3 Effects on Local Residents and Amenities

The Bayston Hill Quarry, proprietors Tarmac Limited, is already covered by stringent guidelines as to how they must minimise the effects of quarrying on local residents and amenities. The northern boundaries of the quarry will not be subject to significant changes; this area already has screen banks. The area that will be affected in terms of the quarry extension is Meole Brace Retail Park and the new site of Shrewsbury Town Football Club. The landscape will need to be enhanced with trees, shrubs and screen banks (British Geological Survey: The Economic Importance of Minerals to the UK, www.mineralsUK.com, accessed 06/04/2011). The quarry at present only affects properties in the Sharpstone Lane area, where the trees at the rear of the houses screen the quarry from view; the quarry is also screened by trees on the opposite quarry boundary. With the proposed extension and relocation of the plant's machinery, new screens are required to shield the properties in Sharpstone Lane (Shropshire County Council, Planning Committee, February 2001 and 7th and 20th March 2007: MS2003/0529/SY and MS2005/1165/SY accessed 04/04/2011).

The proposed quarry extension will be south and south-east of the present quarry edge, where the new major earthworks will take place. These excavations will be more noticeable from Betton Strange Hall and houses on the perimeter of Betton Strange Hamlet. Planted trees, shrubs and hedges will reduce the effects of the quarry extension to these properties. This is an example where the quarry owners have applied the recommendations of Shropshire County Council, Planning Committee, to improve the amenities for local residents (Shropshire County Council, Planning Committee, February 2001 and 7th and 20th March 2007: MS2003/0529/SY and MS2005/1165/SY accessed 04/04/2011).

5.8.2.4 Blasting, Noise and Dust

Blasting times within the proposed quarry extension are limited between the hours 10.00-12.00 and 14.00-16.00; current blasting times at the quarry are 11.00-16.00. So there is a reduction in blasting times from five hour to two hour time slots. This will help the local people and amenities cope with the blasting operation. Tarmac Limited is required to follow procedures and one of these measures is to feedback to the local liaison group, so the blasting operations can be monitored and reviewed regularly. Shropshire County Council Planning Committee required Tarmac Limited under the planning permission to monitor each blast to see if it is within the legal regulations. They are also required to sound sirens; one long siren 15 minutes prior to blasting, three short sirens immediately before blasting commences and two short sirens to sound the 'all clear' after the blasting operations are finished (Shropshire County Council, Planning Committee, February 2001 and 7th and 20th March 2007: MS2003/0529/SY and MS2005/1165/SY accessed 04/04/2011).

5.8.2.5 Noise

Table 5.3 shows the noise levels that must be observed at the different locations. One can see that at Oteley Road and Sharpstone Lane the noise levels are 55.5 dBLaeq, per hour. At Sutton Farm the noise level is 52 dBLaeq, per hour and at Betton Alkmere 48 dBLaeq, per hour. At Bomere Farm the noise level is 49 dBLaeq, per hour. One can see from these figures that there is a difference in permitted noise level of 7 dBLaeq, per hour; this must be because of the difference in relative distances from Bayston Hill Quarries of the road, farms and lane. I know from my site visits to Bayston Hill Quarries, that Sharpstone Lane runs to the entrance of the quarry's main site and processing machinery; it is also close to residential houses and Bayston Village. The close proximity of Bayston Village and Sharpstone Lane to the entrance to Bayston Hill Quarries would account for the greatest noise level being set at 55.0 dBLaeq, per hour by Shropshire County Council Planning Committee. Tarmac Ltd, Bayston Hill Quarries, must obey this requirement of Shropshire County Council Planning Committee; they are also required to monitor and record the noise levels around the said areas and keep their records ready for inspection by the County Council. The wide ranging environmental effect made by the quarrying industry from the 19th century onwards has gradually been mitigated until, under current 21st century legislation, current and future outcomes are monitored and planned in advance under strict planning permission provisions (Shropshire County Council, Planning Permission, Town and County Planning Act 1990, accessed 21/07/2011). The noise levels are shown below:

Table 5.3: Lists: Permitted Noise Levels

Location	Noise Level, dBLaeq, 1hr
1. Oteley Road	55.0
2. Sutton Farm	52.0
3. Betton Alkmere	48.0
4. Bomere Farm	49.0
5. Sharpstone Lane	55.0

(Source: Shropshire County Council, Planning Permission, Town and County Planning Act 1990, accessed 21/07/2011 and Adams, *et al* 1994).

After planning permission has been given, Tarmac Limited have one month to submit a scheme for the monitoring of the noise levels to the Mineral Planning Authority for their approval. At Bayston Hill Quarry the working hours are 07001900 and during this period the permitted noise levels must be adhered to. This also includes any site of relocation of the new plant machinery. Also the free-field noise levels are not to exceed 42.0dB Laeq, per 1 hour at any time, other than during rail loading operations. Rail loading at night shall not go above 45.0dB Laeq, per 1 hour. These noise level controls come under the Structure Plan Policy P59; Minerals Local Plan Policies M3 and M4, designed to protect the local residents and amenities.

5.8.2.6 Dust

Tarmac Limited, have measures in place to control the dust generated by the quarrying operations; nevertheless Shropshire County Council still consider there are opportunities for improvement. The present dust measures at Bayston Hill Quarry are: spraying the stockpiles, washing the wheels on the vehicles and road sweeping Sharpstone Road, which is Bayston Hill Quarry's main access route. The new dust measures could be long-term, or temporary, to cover the processing plant and stockpile areas as well as the internal roads, which enable the transport of quarried stone from place-to-place. Tarmac Limited has in principle agreed with Shropshire County Council, to provide finance for a better air quality monitoring station at Bayston Hill Quarry (Shropshire County Council, Planning Permission, Town and Country Planning Act 1990, accessed 21/07/2011 and Shropshire County Council, Planning Committee, February 2001 and 7th and 20th March 2007: MS2003/0529/SY and MS2005/1165/SY accessed 04/04/2011).

5.8.2.7 Operating Hours

Bayston Hill Quarry operational hours are 0700-1900 and these same hours must be applied to the proposed extension. The Mineral Planning Authority has given limited permission to Tarmac Limited to operate weekends and evenings when necessary, but requests for this have to be submitted in writing to gain approval. At Bayston Hill Quarry there are 38 quarry workers and 35 vehicle owner-drivers who transport the quarried stone across the country. These controls have been put in place to regulate Bayston Hill Quarry and to safeguard the local people and their amenities. This falls under the Structure Plan Policy and Minerals Local Plan Policies

M3 and M4 (Shropshire County Council, Planning Permission, Town and Country Planning Act 1990, accessed 21/07/2011).

5.8.2.8 Landscape Impact

The southern end of the present quarry is where the new extension will be made. This will involve the removal of a line of well-established trees, which delineate the boundary of the present quarry. Land that is presently used for agriculture will be required to site the new processing plant machinery and the new screening banks. These ancient agricultural fields are divided by well-established hedgerows and trees. Tarmac Limited consequently states that they regret the loss of these hedgerows and trees; to compensate for this loss, new planting of trees and hedgerows will take place (Shropshire County Council, Planning Permission, Town and County Act 1990, accessed 21/07/2011 and British Geological Survey: Quarrying and The Environment, www.bgs.ac.uk/mendips/aggregates/ environment, accessed 20/07/2011).

5.8.2.9 Ecological Considerations including effects on Sensitive Sites and Animals Species

Shropshire County Council asked for an ecological survey to be undertaken as part of the environmental statement. Whilst The Mineral Planning Authority had given their approval for the quarry extension, as part of the development works, within three months of the first approval, a further plan for the proposed removal of hedgerows, trees, shrubs, a pond and other habitats needed to be submitted and additionally approved by this authority. However, there are birds of prey such as peregrine falcons, known to be residing in the existing quarry (Shropshire County Council, Planning Permission, Town and County Act 1990, accessed 21/07/2011 and Shropshire County Council, Planning Committee, February 2001 and 7th and 20th March 2007: MS2003/0529/SY and MS2005/1165/SY accessed 04/04/2011).

This ecological survey discovered a badger's sett on the proposed quarry site; Tarmac Limited responded by stating that this was only one single badger hole and its removal would not affect the local badger population. Nevertheless monitoring of this badger sett would be carried out. In the Tarmac Limited application, a recording of the ordinary species of fauna and flora would be carried out for this area, so that the species removed will be replaced in the future restoration schemes (Shropshire County Council, Planning Committee, February 2001 and 7th and 20th

March 2007: MS2003/0529/SY and MS2005/1165/SY accessed 04/04/2011 and British Geological Survey: Quarrying and The Environment, www.bgs.ac.uk/mendips/aggregates/environment, accessed 20/07/2011).

The County's Ecologist requested that a further survey be conducted, after changes had been made in both policy and guidance regarding protection of species. This would be a more in depth survey which would include the badgers' sett and investigate whether or not species such as: bats, great crested newts, dormice, reptiles and breeding birds have been established at the site. The survey indicates that great care should be taken when trying to identify bats present. All these species are protected under The Natural Environment and Rural Communities Act 2006 (Lobo Ecology-Species and Habitat, www.loboecology.co.uk/species, accessed 21/07/2011 and The Natural Environment and Rural Communities Act 2006, www.naturalengland.org.uk/oruwork/conservation/biodiversity/protectandnama, accessed 21/072011).

This survey concluded that the quarry extension and associated developments could take place without any adverse effects to the protected species. However the surveys did drawn attention to the possible time delay between the granting of planning permission and start of the quarry extension. This time delay could provide an opportunity for any protected species to inadvertently re-establish itself at the site of the quarry extension, so that it will be necessary for Tarmac Limited to assure Shropshire County Council's Planning Committee that no such re-colonising has taken place before planning permission can be granted. In addition, a method statement must be submitted to the County Council before each phase of the work commences. The report suggested that Tarmac Limited should have a licensed bat handler on hand when the trees are felled in the spring or autumn in case any bats are residing in the trees. Tarmac Limited stated that they were willing to adopt these requirements. As before, attention was drawn to the short-term and long-term restoration plans for the agricultural fields and present woodland after the quarry extension has been exhausted (Shropshire Town and Country Planning Act 1990, accessed 21/07/2011, Shropshire County Council, Planning Committee, February 2001 and 7th and 20th March 2007: MS2003/ 0529/SY and MS2005/1165/SY accessed 04/04/2011 and The Natural Environment and Rural Communities Act 2006, www.naturalengland.org.uk/oruwork/ conservation/biodiversity/protectandnama, accessed 21/07/ 2011).

5.8.2.10 Public and Access Rights of Way

At Bayston Hill Quarry there is a public right of way in the vicinity of the works. It runs from Sharpstone Lane up past the processing plant and along the southern ridge of the present quarry; it then continues towards Sutton Farm, connecting with Oteley Road. In Tarmac Limited's quarry extension application, they propose that the present public right of way should be diverted to the southern edge of the quarry. This would allow the new quarry, processing plant and screening banks to be developed. During my site visit to Bayston Hill quarry on the 29/03/2011, I walked along the re-sited public right of way which provides an excellent view of the former quarries, now backfilled, as well as the present working quarry. Tarmac Limited also intends to place interpretation and information boards which would be placed at prominent positions along the route. Landscaping around the public right of way has already started, but is at an early stage. The trees and shrubs, when established, will screen the quarry from sight and provide an excellent habitat for wildlife; additionally, there is a magnificent panoramic view (Shropshire County Council, Planning Committee, 7th and 20th March 2007: MS2003/0529/SY and MS2005/1165/SY accessed 04/04/2011, The Natural Environment and Rural Communities Act 2006 and Galloway, Bayston Hill site visit 29/03/2001).

5.8.2.11 The Effects on Agricultural Land

The Department of Environment, Food and Rural Affects (DEFRA) raised concerns about the removal of more than 30 hectares of prime agricultural land when Tarmac Limited first submitted their application. The stripping of the land would allow for the overburden and waste from the quarry extension to be stored. DEFRA's concern was that the landscape would not be restored to its original agricultural use, but to a combination of grassland and broadleaf woodland. The new quarry extension will provide an additional 30 million tonnes of greywacke, a highly quality mineral, plus a new plant processing and stockyard site. DEFRA, after studying all the facts, still wish that a percentage of the quarried land should be returned to arable, agricultural use. In addition, DEFRA request that the topsoil and subsoil, once removed, should be carefully handled and stored. The planning policies cover landscape restoration and require that it is done to a high standard (The Department for Environment, Food and Rural Affairs, www.defra.gov.uk, accessed 22/07/2011 and Shropshire County Council, Planning Committee, 7th and 20th March 2007: MS2003/0529/SY and MS2005/1165/SY accessed 04/04/2011).

5.8.2.12 The Archaeology and Cultural Heritage

Tarmac Limited have been informed by Shropshire County Council that before any work on the quarry extension can commence, they will need written letters from both The Mineral Authority and The Council's Historic Environment Officer. It is quite possible that a watching brief will be required to supervise the removal of the topsoil and subsoil from the new quarry site. This is also to ensure that, if need be, an archaeological investigation and evaluation can be carried out, and thus any new finds be recorded (Shropshire Town and Country Planning Act 1990, accessed 21/07/2011 and Shropshire County Council, Planning Committee, 7th and 20th March 2007: MS2003/0529/SY and MS2005/1165/SY accessed 04/04/2011).

Archaeology, whether buried or above ground, can include anything that provides evidence of the impact of past human activity upon the landscape. This is important for both the archaeological, historical and environmental records. Greywacke is a limited resource in Britain so every site, similar to Bayston Hill Quarry, is very important. Though saying this, archaeological remains are also an irreplaceable resource and can occur anywhere. Archaeological artefacts, in many cases, are very fragile and vulnerable to damage and destruction. They are not all of equal importance, but all archaeological remains have some value; therefore their significance can only be established by proper investigation, which involves thorough recording of find-sites and protection of objects, because once archaeological artefacts have been destroyed they cannot be replaced. This is why both the planning authorities and quarry developers should work together to ensure that all archaeological artefacts are properly investigated and recorded. With this understanding in place both planning authorities and quarry developers can achieve their separate objectives (Mineral Extraction and Archaeology: A Practice guide, accessed 02/12/2011 and Mineral Extraction and The Historic Environment, accessed 02/12/2011).

5.8.2.13 Restoration of Quarry Site and Aftercare

After 2060, when all the quarries at Bayston Hill will have ceased to operate, Tarmac Limited the present owners of the land will have needed to create grassland and woodland habitats, in order to return the site into an area of the greatest biodiversity. The restoration of other Bayston Hill former quarries, now backfilled, has yet to be commenced; nevertheless the planting of trees and shrubs on both sides of the public right of way is a very impressive start. The new quarry on the southern edge of the present quarry will be screened by trees and shrubs; when established these will be an everlasting feature in the landscape. Once the removal of the topsoil and subsoil has been completed, the construction of the new processing plant can start and become operational. This will then allow for the restoration of the present processing plant and stockyard located at the top of Sharpstone Lane, which can be seen from both north and south Bayston Village, as well as from the A49 approach road to the village. The restoration of the site with grassland, woodland and shrubs will eventually restore the natural beauty of the place (British Aggregates Association, www.british-aggregates.co.uk/about/restoratio, accessed 22/07/2011, Restoration Success at Tarmac Quarry, www.tarmac.co.uk/aboutus/news/restorationsuccessattarmac, accessed 2/07/2011 and Shropshire County Council, Planning Committee, 20th March 2007, MS2003/0529/ SY, accessed 04/02/2011).

The present and future options for restoration of Bayston Hill Quarry are limited, because of the depth of excavation. When fully developed and operational, the approximate amount of greywacke mineral excavated from the future quarry would be in the region of 1,200 million tonnes per year. Because of the amount of stone removed there will be insufficient quarry waste, or overburden, to use as backfill to bring the quarry site back to its original level. There are proposals to incorporate a water feature within the base of the quarry because of the eventual final depth. The former plant site will become a species rich grassland area. The upper faces and benches will be selectively backfilled to encourage recolonization and regeneration of the former quarry site. The Planning Committee have made it clear from the start that no bulk or other waste materials can be used in backfilling the quarry, or in levelling, or restoration of any part of the quarry site. There is some leeway for the use of topsoil in the restoration of Bayston Hill Quarry, but prior authorization must be obtained from The Mineral Planning Authority (Shropshire Town and Country Planning Act 1990, accessed 21/07/2011 and Shropshire County Council, Planning Committee, 7th and 20th March 2007: MS2003/0529/SY and MS2005/1165/SY accessed 04/04/2011).

5.8.2.14 Aftercare

The Mineral Planning Authority has stated that they will require a plan of action in writing for the aftercare of the former quarry site for their approval within 3 months of the restoration work being completed. When the aftercare scheme has been approved, conditions will be specified for the steps needed to bring the land back to the required standards for its subsequent use as:

- Agricultural use: as far as it is possible the land should be restored to its former agricultural usage.
- Woodlands and amenities: the land should be restored to conditions that are capable of sustaining trees, shrubs, grassland areas and water features. Also the rocks faces must be stabilised to maintain the angles and benches, which in turn can be planted with trees, shrubs and other plants, in order to promote a habitat suitable for recolonization by wildlife. These areas must be to a standard that incorporate safety, particularly of the rock faces.

The Mineral Planning Authority requires a written statement, submitted at the end of each year, outlining the work that has been carried out that year, as well as work that will be carried out in the next year. A minimum five year aftercare period for the former quarry land is required after it has been restored to agricultural use. Areas such as woodlands, quarry benches, quarry faces, water features, and habitat spaces will need a more substantial period of aftercare. The Mineral Planning Authority estimate that after the restoration of the land, once occupied by the former quarry, a maximum period of 25 years aftercare will be required (Shropshire Town and Country Planning Act 1990, accessed 21/07/2011 and Shropshire County Council Planning Committee, Proposals for Planned Restoration, accessed 16/03/2011).

5.9 Archaeology

5.9.1 Bayston Hill

Archaeological remains are irreplaceable and a limited resource. Both Bayston Hill and Brown Clee Hill Quarries have, or had, archaeological remains which are important to the archaeological and historical records. This is why we need legislation to protect such remains when discovered, either when quarrying is taking place, or when land is prepared for access to roads and railway-lines to and from the quarries.

When in 2009, Tarmac Limited wanted to expand to the east of the present greywacke quarry the Planning Committee of Shropshire County Council required an archaeological investigation to be carried out, as a condition of consent for any quarry expansion. This was to be undertaken to evaluate the known Roman road that crosses Bayston Hill from the south-east. Below is a picture of the Bayston Hill present quarry and proposed extension:

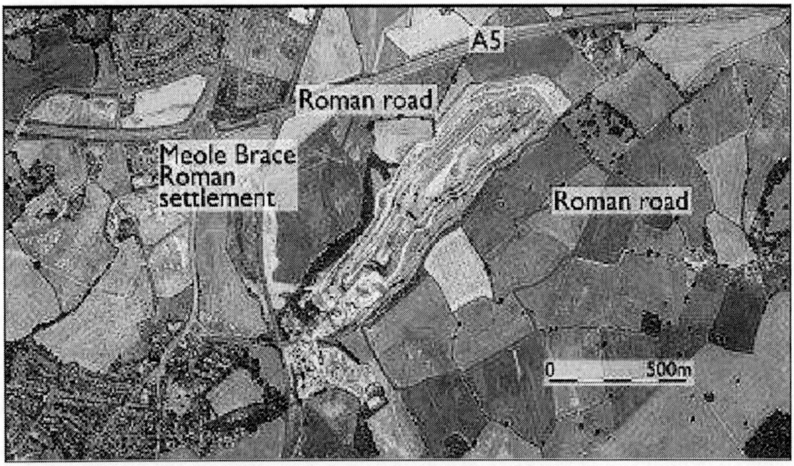

Source: British Archaeology 120, September/October 2011, www.archaeologyuk.org, accessed 09/09/2014.

As one can see from the above picture, a Roman road did exist, at one time called the Portway. It is suggested that the route might have connected with Haughhmond Hill, Ebury hillforts to the north-east and with the Burgs at Bayston Hill, then continuing south-west to the Long Mynd (British Archaeology, www.archaeologyuk.org, accessed 9/09/201 and Malim, *et al,* 2011).

Other sites of archaeological and historical importance discovered on or around Bayston Hill in the 1960s were the Neolithic, Bronze Age ditches and cemetery. Also, north-east of the Burgs on Bayston Hill are Bronze Age barrows and late Bronze Age, or early Iron Age, field ditches; 800 metres north of this same site are rectilinear enclosures (Malim, *et al,* 2011).

5.9.2 Brown Clee Hill

The remains of several hillforts are located around the Clee Hills; Nordy Bank is the only intact fort remaining of the original three Iron Age hillforts on Brown Clee Hill. Abdon and Clee Burfs were largely destroyed by quarrying activities in the 19th and 20th Centuries. These particular hillforts were built on Brown Clee, the highest hill in Shropshire, and we can now never appreciate the special significance they held at that time. From 1907 to 1908 The Congress of Archaeology Societies, now known as the Council for British Archaeology, expressed their concern about the quarrying activities on Brown Clee Hill. Their concern was for the possible loss of the early Neolithic, Bronze Age sites that could be consumed by quarrying. The Congress of Archaeology Societies requested that Brown Clee Hill Stone Company suspend all quarrying activities while an archaeological evaluation took place; this request was refused. At this time no legislation was in existence to protect archaeological sites. This is why the Mineral and Archaeology: Practice Guide, 2006 is essential today (Bryan, 2006).

5.10 Discussion

To this day, Bayston Hill Quarry is still producing greywacke stone for road construction; however Brown Clee Hill Quarry dolerite, also used in road construction, operations ceased in the early 20th century. Researching the two sites was very different, because Bayston Hill Quarry, owned by Tarmac Limited, would not allow access to their archived records. This was different at Brown Clee Hill as all its records are now in the Shropshire Archives and can be viewed by appointment. However, there were other ways around researching Bayston Hill Quarry. Local Councillor Ted Clarke and the Bayston Hill Village Local Council Office kindly lent me the complete parish records with the planning applications that appertain to Bayston Hill (Ted Clarke, per comm, 03/03/2011). Extensive research at Shropshire Archives did provide a little information on previous owners of Bayston Hill Quarry and a map of 1789, but apart from that very little recorded documentation is available on Bayston Hill Quarry. Bayston Hill Village library had no information on the local quarry.

Bayston Hill Quarry is a unique site for the mineral "greywacke"; this was evident when a government-sponsored report was carried out at Bayston Hill. The report defined the quarry as being among the top four producers of greywacke in the country; thus confirming that it is a major producer of High Specification Aggregates. With such a report one can suggest that it is very unlikely that any planning application would be turned down, however it would be subject to stringent environmental restrictions. The research highlighted concerns about environmental issues such as road access and noise from Bayston Hill Quarry. Road access had been a major factor in the Wenlock Edge Quarries, contributing to the reason why Lea Quarry was mothballed. Quarrying has been carried out for many centuries in one form or another, but more recently concerns about the protection of special species fauna and flora, which could be lost, are now considered in planning applications. The potential loss of archaeological remains, such as Neolithic, Bronze Age sites and Roman roads are also reviewed before approval of planning applications.

The quarries on Brown Clee Hill that once produced dolerite for tramway setts and road construction today has now been reclaimed by nature; however the scars of its industrial past are still evident in the 21st century. These scars are on the landscape in the forms of quarries and derelict buildings. Brown Clee Hill today is used by people who like walking the hill to look at fauna and flora on the former quarry landscape; care should be taken as some of the old quarries are now deep pools of water. On Brown Clee Hill sites of archaeological interest were destroyed by quarrying when there was no legislation to protect them. So this is why it is important that Bayston Hill Quarry, although its mineral is of national significance to the country, must now protect the Sites of Special Scientific, archaeological and environmental interest before any further, future quarrying takes place.

5.11 Conclusion

Bayston Hill Quarry is still operating today and its greywacke is still in great demand; Brown Clee Quarries, which quarried dolerite, has been closed for some considerable time. The location of Brown Clee Quarry, having poor rail and road access, was one of the main reasons for its demise. Bayston Hill Quarry had used the railway to transport its quarried stone until the 1950s when the railway sidings linking the quarry to the main railway-line ceased operating. In 2007, Tarmac Limited made an application to Shropshire County Council for the reopening of the railway sidings, which was granted. They reconsidered using rail transport because of the economics associated with road haulage. The new railway sidings linked Bayston Hill Quarry to the main Shrewsbury to Cardiff line, formerly the Newport to Gwent railway-line.

Brown Clee Hill Quarry, because of its geological location, made a main connection with major rail transport impossible; even though they had The Cleobury Mortimer and Ditton Priors Light Railway, it only lasted for some 13 years. Dolerite was not an inexhaustible resource on Brown Clee Hill and although the company tried to diversify into prefabricated building construction, the writing was on the wall for the Brown Clee Hill Company having to close for good.

Another difference between Bayston Hill and Brown Clee Quarries was road access; Bayston Hill is conveniently located near to the A49, which runs straight through Bayston village and near to the A5 and a network of motorways. However, Brown Clee Hill is not located near a main road, only the minor road B4364; so one can see that the geological location of a quarry still plays a significant role, even up until the 21st century. Location also affected the workforce of the quarries; getting to Bayston Hill Quarry was considerably easier that getting to work at Brown Clee Hill. At Brown Clee, the workforce had to walk up the hill to its summit even before starting work. If a quarryman was lucky enough to be provided with a company house the regulations laid down by the quarry owners were very restrictive; these included working double shifts in the summer to accommodate being laid off in the winter period.

The future for Bayston Hill and Haughmond Quarries is not straight forward because of all the concerns about the environment. During negotiations with Shropshire County Council they were informed that they must not work Bayston Hill and Haughmond Hill Quarries simultaneously. In 2060 when all quarrying operations must have ceased at Bayston Hill, only then can Tarmac Limited apply to Shropshire County Council for permission to extend the quarrying at Haughmond Hill, but not before.

From 2015 Bayston Hill Quarry has 45 years to run until January 2060 before it returns to nature similar to Brown Clee Hill Quarry. The life of Haughmond Quarries continues into the future, but how will these pressures and concerns impact on environmental issues at this quarry in the future?

CHAPTER 6

INDUSTRIAL: LILLESHALL, WENLOCK EDGE QUARRIES

6.1 Introduction

This chapter chronicles the development of the limestone industry at Lilleshall and Wenlock Edge Quarries, Shropshire, during the 19th and 20th centuries. However it will investigate the earliest beginnings of the Lilleshall and Wenlock Quarries to form a complete analysis leading up to the period under study. It will examine the geological, archaeological and historical features of the two sites.

6.2 Geology of the Lilleshall SJ734165 & Wenlock Areas SJ612996

The County of Shropshire has diverse geological formations and is rich in limestone, having within it both minerals from the Silurian and Carboniferous periods. The Silurian strata were formed 435 to 410 million years ago when Shropshire lay under the Iapetus Ocean and has been the subject of many important geological studies starting with Lapworth. The Silurian geological system of Shropshire is formed from mudstone alternating with limestone. This extends in a south-westerly direction from Ironbridge Gorge to Wenlock Edge; continuing to the south-east as far as Carven Arms (Lapworth *et al,* 1910 and Toghill 2006, pp. 123-52).

Lilleshall Hill has been formed from the rhyolitic tuffs of the pre-Cambrian volcanic mass rising up out of the east Shropshire landscape to create the northern limit of the eastern Uriconian Outcrop; among the oldest volcanic intrusions in Europe (Toghill, 2006, 47). The limestone in and around Lilleshall accumulated against these deposits. The Lower Carboniferous limestone, which forms the long narrow outcrop at Lilleshall Hill, is split by a number of major faults named: Lightmoor, Wrekin and Brockton. By the 1920s quarrying operation had ceased; consequently the old quarries were inaccessible, either overgrown or flooded and reclaimed by nature; additionally, the mines had been flooded and therefore were unreachable (Ove Arup *et al*, 1987).

Geology at Wenlock Edge is formed from a double scarp running north-east to south-west. The Wenlock Ridge continues parallel to the parish boundaries in the townships of Farley, Wyke, Bradley and Presthope. There are two strata: one, The Lower Wenlock Limestone, which has very variable characteristics, and secondly, The Aymestry Limestone, a much harder rock, which is found in the higher escarpment. Weathering has played its part on the Silurian Limestone Formation at Wenlock Edge, which is covered and underlain by shale. These beds dip to the south-east and there is a gentle slope in this direction. The Aymestry Limestone runs parallel to the Wenlock scarp, approximately one mile to the east (Toghill, 2006 and Silveter, *et al*, 1989).

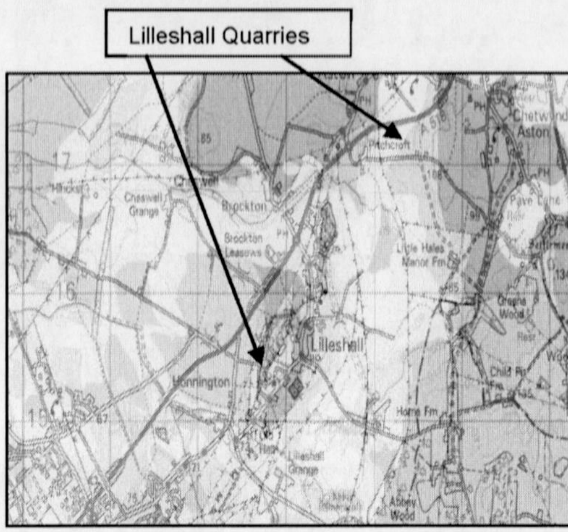

Figure 6.1: Geological map of Lilleshall Hill Source: Geology Roam. Digimap.edina.ac.uk/geologyroam accessed 24/04/2012).

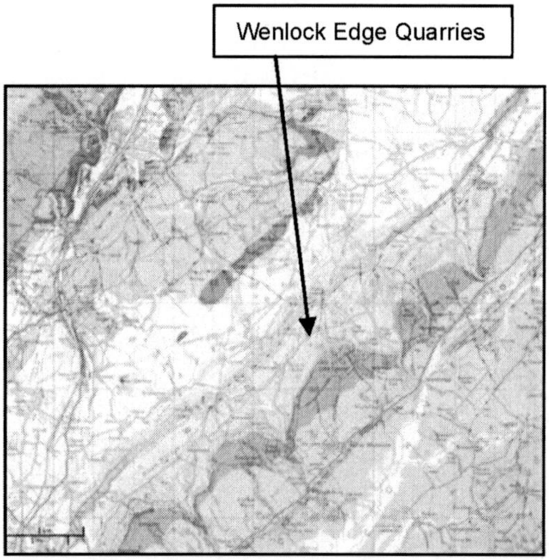

Figure 6.2: Geological map of Wenlock Edge. (Source: Geology Roam. Digimap.edina.ac.uk/geologyroam, accessed 24/04/2012).

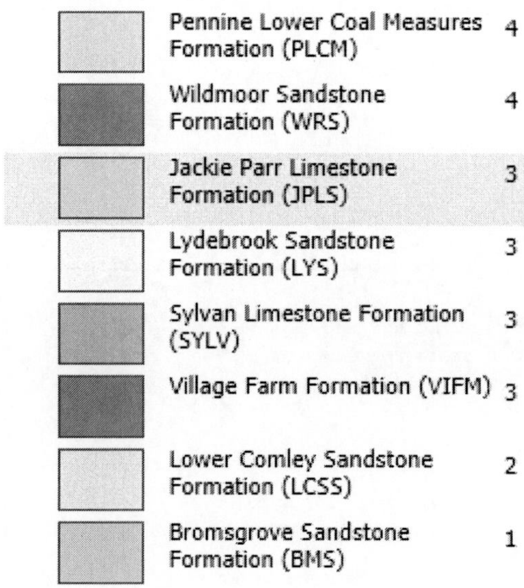

Figure 6.3: Geological Legend Lilleshall (Source: Geology Roam. Digimap.edina.ac.uk/geologyroam, accessed 24/04/2012).

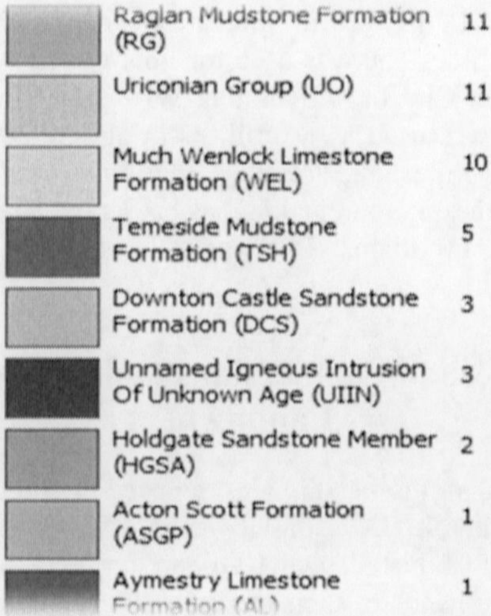

Figure 6.4: Geological Legend Wenlock Edge (Source: Geology Roam. Digimap.edina.ac.uk/geologyroam, accessed 24/04/2012).

6.3 Lilleshall and Wenlock Edge Limestone Qualities

Lilleshall and Wenlock Edge Limestones are of the highest qualities; especially so the Wenlock Edge Limestone at Lea Quarry which is unique and for this reason was mothballed in 2007. At Lilleshall and Steeraway there are outcrops of carboniferous limestone. The Visean strata containing white limestone are part of the rock formation from the Lower Carboniferous period. The Lilleshall area has two types of limestone, one the white limestone which is of the highest quality, was used as flux in blast furnaces. The other grey limestone was used in the construction of buildings and was applied to the land being used in agriculture (Cossey, *et al,* 2004).

Wenlock Edge Limestone, in sedimentary terms, is a comparatively hard and resistant rock, lying between the two softer rocks of the Wenlock Shale to the northwest and Ludlow Shale to the south-east. Wenlock Limestone has bluey-grey colouration and is found in calcareous beds, approximately 35 metres thick. Southwest of Easthope, where the reefs are not present, the limestone thickens to approximately 137 metres (Toghill, 2006 and Williams, 1997). Wenlock Edge limestone has been used from the earliest days as a building stone locally as can be seen in the remaining

Priory Walls and the Corn Exchange in Wenlock Town. More significantly, the limestone was used for lime mortar, especially as can be seen in the Roman City of Viriconium, Wroxeter. Since before medieval times, limestone was used as a fertiliser in agriculture (Williams, 1997). However, as with Lilleshall, Wenlock Edge limestone was also used as a flux in blast furnaces, in road making, as bricks or slabs and in the makeup of artificial stone (Williams, 1997, and The Quarry Managers' Journal, 1968).

6.4 Lilleshall Partnerships and Wenlock Edge Landowners

In 1802 the Lilleshall Partnership was replaced by the Lilleshall Company; at this time Granville Leveson-Gower 1st Marquis of Stafford's was approaching his 80th birthday, the Gilbert brothers were dead and their shares were in the hands of their executors. Partners in the new Lilleshall Company were Lord Granville Leveson-Gower Ist Earl of Granville son of the 1st Marquis of Stafford's third marriage and other entrepreneurs with landed interests, capitalist and technical skills (Adams, 2007 and Gales, *et al,* 1979). During February and March 1802, Lord Granville Leveson-Gower obtained the Gilbert brothers' shares in the former Lilleshall Partnership. In April 1802 Lord Leveson-Gower's father, the 1st Marquis of Stafford, assigned him a further share in his iron and limestone workings in Lilleshall and adjoining parishes (Ironbridge Gorge Museum Trust, Lilleshall Collection, 1998.320, (DLIL/3) 174, 176 Assignment of Shares and 1998.320, (DLIL/3) 178 Articles of Co-Partnership, accessed 05/08/2009).

Table 6.1 shows the articles of the Lilleshall Company which were signed on 24th June 1802, with Lord Granville Leveson-Gower taking 50% of the shares, John Bishton 5/16 shares, John Onion, James Birch and William Philips 1/16 shares each (Ironbridge Gorge Museum Trust, Lilleshall Collection, 1998.320, (DLIL/3) 174, 176 Assignment of Shares and 1998.320, (DLIL/3) 178 Articles of Co-Partnership, accessed 05/08/2009). The next column records the agreement signed between Earl Gower and Mr James Coles for any restoration work that might be required to his landscape should the Donnington Wood Canal close. A document was found referring to the labour used in the construction of the Donnington Wood Canal; and the time it took to construct which was from 1765 to 1767 when it was completed. The last column refers to Pitchcroft Limestone Mine, which is referred to later in my book. It again records covenants of articles regarding the partnership of the Lilleshall Company.

The Donnington Wood Canal was finally abandoned in 1904, however on my field visit; evidence of the canal was still visible. The share issues show the complex workings of Lord Granville Leveson-Gower and how he created the company; by this time, in the late 18th century, the Gilbert brothers had died, so new partners were needed to create a new company (Ironbridge Gorge Museum Trust, Lilleshall Collection, 1998.320, (DLIL/3) 174, 176 Assignment of Shares and 1998.320, (DLIL/3) 178 Articles of Co-Partnership, accessed 05/08/2009). Below is a table showing the details:

Table 6.1: Contents of Lilleshall Iron & Steel Company documents held by IGMT

Document reference no:	Records found:
1998. 320 (DLIL/3) 174,176 Assignment of Shares and 1998. 320 (DLIL/3) 178 and Articles of Co-Partnership	This document discusses the allocation of shares in the new Lilleshall Company of 1802.
1997. 320 (DLIL/3) 2202/38	Earl Gower agreement with Mr James Coles for any damage or restoration work to his landscape, if Donnington Wood Canal closes.
1997, (DLIL/3) 2202/5	Labour used on construction of Donnington Wood Canal and time it took to build it.
1998.320 (DLIL/3) 556a 1998. 320 (DLIL/3) 237. Covenants in Articles of Partnership	Evidence concerning Pitchcroft Limestone Mine. This document discusses the partnership of William Horton and Lord Gower in the Lilleshall Company.

(Ironbridge Gorge Museum Trust, Lilleshall Collection, 1998.320, (DLIL/3) 174, 176 Assignment of Shares and 1998.320, (DLIL/3) 178 Articles of Co-Partnership, accessed 05/08/2009).

Three or four years later ownership of the shares in the Lilleshall Company changed hands when the shares of James Birch and the Onion family were assigned over to John Bishton senior; however by 1806 John Bishton was dead. John Bishton junior succeeded his father and in turn he, by his brother George Bishton, by joining the Lilleshall Company. In the case of John junior and George Bishton, their late father's executors negotiated a new partnership with Lord Granville Leveson-Gower, whose Snedshill mines and ironworks would now be brought under the control of the Lilleshall Company (I.B.G.M Trust, Lilleshall Collection, 1998.320 (DLIL/3) 1992 memorandum of Agreement and 1998.320 (DLIL/3) 210a and 210b, accessed 05/08/2009). In 1813 William Horton joined the Lilleshall Company partnership; previously he was connected with William Reynolds and The Coalbrookdale Company. William Horton and his son John were prominent in the dealings of the Lilleshall Company for most of the 19th century (I.B.G.M Trust, Lilleshall Collection, 1998.320(DLIL/3) 237, accessed 05/08/2009 and Staffordshire Record Office, Sutherland Collection, D593/B/2/7/22, accessed10/11/2009).

6.4.1 Wenlock Edge Landowners

As early as 1714 limestone quarrying on a small scale had existed on Farley Common, but under the 1818 Enclosure Act some 102 acres of Farley Common was enclosed, thus bringing an end to the small limestone quarry operators. This therefore attracted the attention of various major land owners. Since the late 1700s, Farley Farm was in the ownership of Richard Reynolds an ironmaster. In 1818, Cecil Weld-Forester, after the Enclosure Act, received woodland on Farley common. In 1841, the principal landowners in the townships of Farley, Wyke and Bradley, within the Parish of Much Wenlock, were Abraham Darby, Sir E.E Wynn and Lord Forester. In 1848, Lord Forester owned 194 acres of land in and around the Much Wenlock town area, following this, in 1858 he bought 827 acres from Sir Watkin Williams Wynn, so owning over half of the Much Wenlock manorial estate (Shropshire Archives, 1224, box 108 deeds and documents 1681/42/16, accessed 23/04/2014). In 1858, Caleb Harper bought the woodland on Farley common from Cecil Weld-Forester; also in 1889-1891 he acquired Reynolds and Darby's leaseholds on Farley Common. When in 1905 Caleb Harper died, he left 260 acres in Farley and Bradley Common to his nephew R. H. Trevor, who in turn left these lands to his son Hubert Trevor. On September 25th 1918, Lord Forester sold his Dothill estate by auction, which included manorial lands in Little Wenlock together with 1,325 acres of outlying lands of The Sparrow's Church estate. Lord Forester's estate at Willey still remains in the Forester family. In 1930, Lord Forester bought back the 260 acres of land at Farley and Bradley Common (Shropshire Archive, Shropshire Chronicle, 1918, accessed 23/04/2010, Shropshire Archives, document no, 1206/99, accessed 23/04/2010 and Bagshaw, 1851). Farley Quarry is located north of Much Wenlock and today is the only working quarry left in the area, but with many restrictions restraining its future development, mainly with respect to environment and agricultural constraints (Mineral Extraction on Wenlock Edge, 1979). The Wenlock Edge area is now under the stewardship of The National Trust (The National Trust, www.nationaltrust.org.uk/Wenlock-edge, accessed 16/09/2016).

In 1802 the Lilleshall Partnership was replaced by the Lilleshall Company; at this time Granville Leveson-Gower 1st Marquis of Stafford's was approaching his 80th birthday, the Gilbert brothers were dead and their shares were in the hands of the executors. Partners in the new Lilleshall Company were Lord Granville Leveson-Gower Ist Earl of Granville son of the 1st Marquis of Stafford's third marriage and other entrepreneurs with landed interests, capitalist and technical skills (Adams, 2007 and Gales, *et al,* 1979). During February and March 1802, Lord Granville Leveson-Gower obtained the Gilbert brothers' shares in the former Lilleshall Partnership. In April 1802 Lord Leveson-Gower's father, the 1st Marquis of Stafford, assigned him a further share in his iron and limestone workings in Lilleshall Company. The articles of the Lilleshall Company (partnership) were signed on 24th June 1802, with Lord Granville Leveson-Gower taking 50% of the shares, John Bishton 5/16 share, John Onion, James Birch and William Philips 1/16 each (Ironbridge Gorge Museum Trust, Lilleshall Collection, 1998.320, (DLIL/3) 174, 176 Assignment of Shares and 1998.320, (DLIL/3) 178 Articles of Co-Partnership, accessed 05/08/2009).

Three or four years later ownership of the shares in the Lilleshall Company changed hands when the shares of James Birch and the Onion family were assigned over to John Bishton senior; however by 1806 John Bishton was dead. John Bishton junior succeeded his father and in turn he, by his brother George Bishton, by joining the Lilleshall Company. In the case of John junior and George Bishton, their late father's executors negotiated a new partnership with Lord Granville Leveson-Gower, whose Snedshill mines and ironworks would now be brought under the control of the Lilleshall Company (I.B.G.M Trust, Lilleshall Collection, 1998.320 (DLIL/3) 1992 memorandum of Agreement and 1998.320 (DLIL/3) 210a and 210b, accessed 05/08/2009). In 1813 William Horton joined the Lilleshall Company partnership; previously he was connected with William Reynolds and The Coalbrookdale Company. William Horton and his son John were prominent in the dealings of the Lilleshall Company for most of the 19th century (I.B.G.M Trust, Lilleshall Collection, 1998.320(DLIL/3) 237, accessed 05/08/2009 and Staffordshire Record Office, Sutherland Collection, D593/B/2/7/22, accessed10/11/2009).

6.5 Lilleshall and Wenlock Edge, 19th and 20th Century Quarrying Evolution

An evaluation of 19th and 20th century maps of Lilleshall and Wenlock Edge area. On the 1900 map of the Lilleshall area there are three old quarry sites recorded. North of the Lilleshall Village is the first old quarries site, the second old quarries site is to the east and the third old quarries site is west of the village. In addition The Lilleshall Barracks House can be clearly seen on the 1900 map; as can the old routes of the former canal and disused tow path. By the former canal and disused tow path are two old shafts. The short stretch of canal was used to transport limestone quarried from the connecting quarries. To the same map to the west of the village can be seen the Shropshire Union Railway Line (Digimap, www.ac.uk/ancientroam, accessed 25/03/2014).

The 1920 map Lillishall shows the Shropshire Union Railway Line, Barrack Houses, Old shafts, Canal Branch and Tow Paths. Again the 1950 map shows everything but the Barrack Houses. Again the Lilleshall map of 1960 shows the sites of the old quarries, but they are not named. The Shropshire Union Railway and the former canal are still visible and four former old shafts. The 1980 map of Lilleshall shows two disused kilns sites in the former quarries. The former canal branch is now indicated by dotted-lines and the Shropshire Union Railway is marked as dismantled. The 1990 map of Lilleshall show disused kilns, one old shaft and disused canal (Digimap, www.ac.uk/ancientroam, accessed 25/03/2014). Below are the maps of Lilleshall:

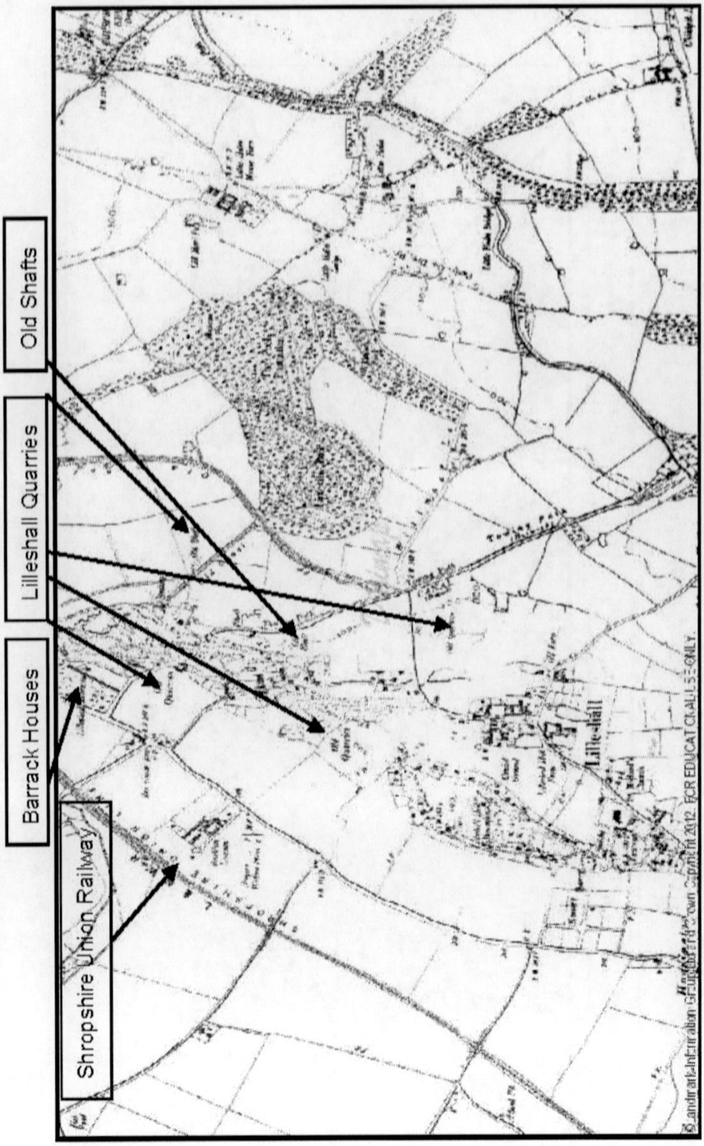

Figure 6.5: Lilleshall 1900. (Source: Digimap, www.ac.uk/ancientroam, accessed 25/03/2014).

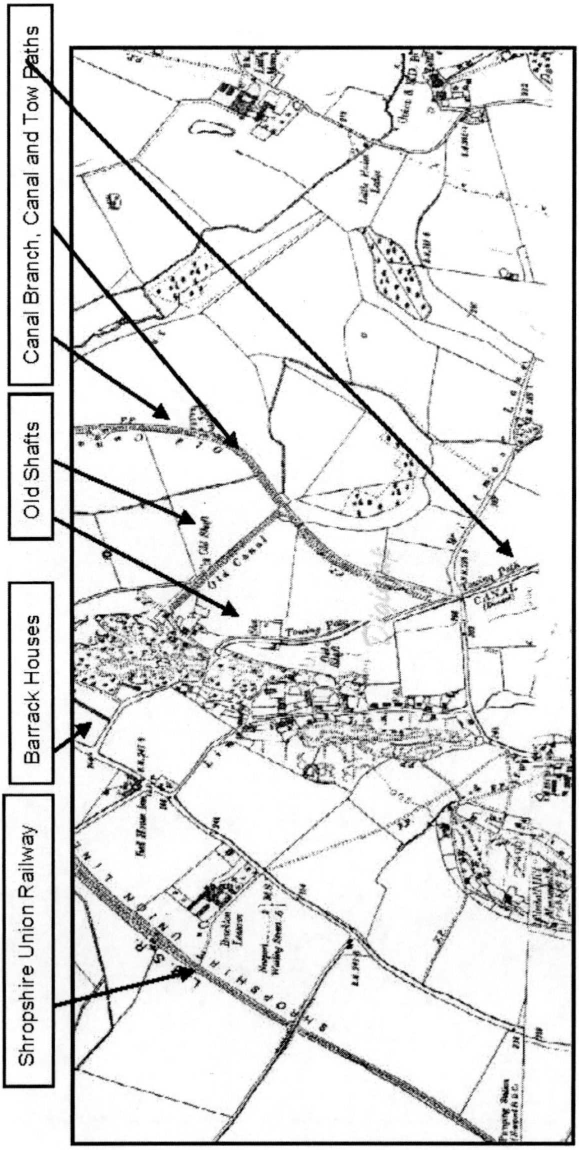

Figure 6.6: Lilleshall 1920. (Source: Digimap, www.ac.uk/ancientroam, accessed 25/03/2014).

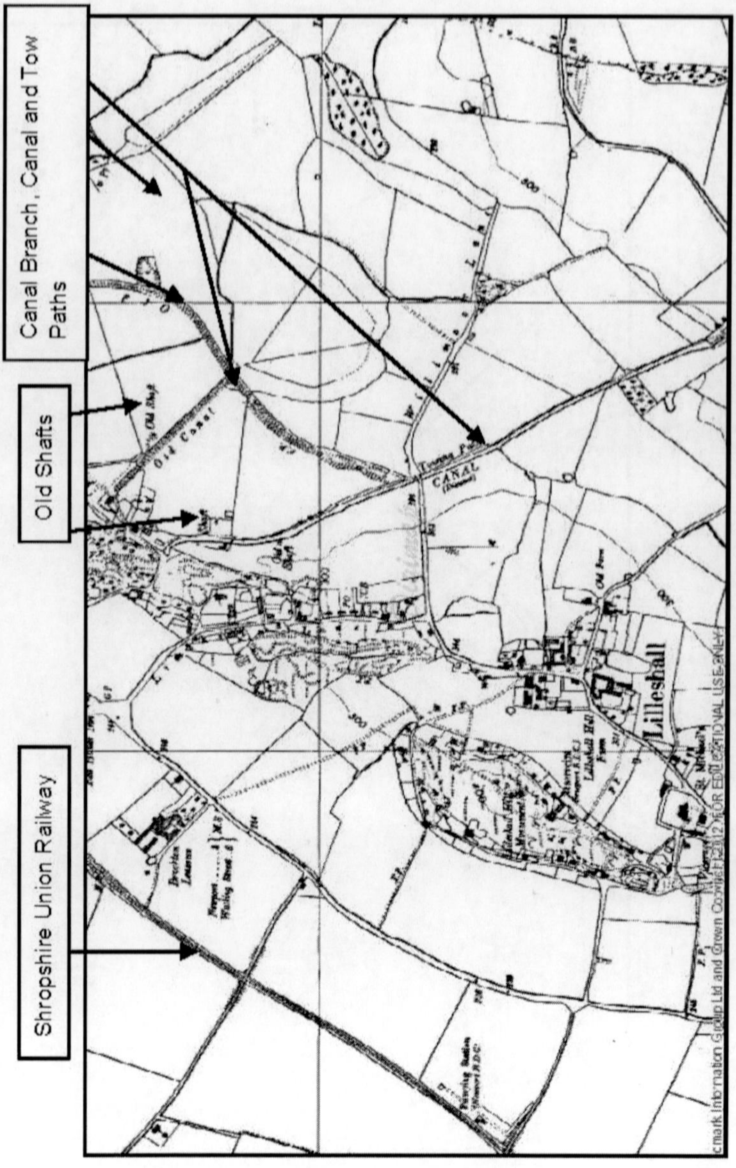

Figure 6.7: Lilleshall 1950. (Source: Digimap, www.ac.uk/ancientroam, accessed 25/03/2014).

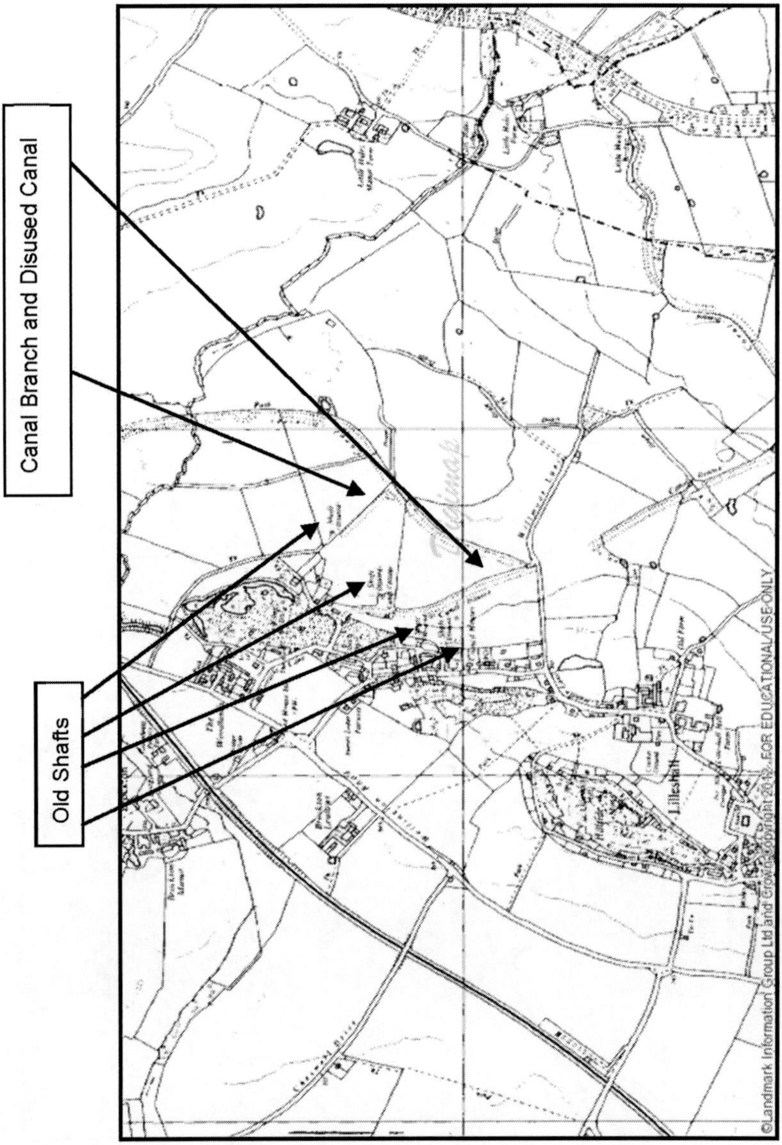

Figure 6.8: Lilleshall 1960. (Source: Digimap, www.ac.uk/ancientroam, accessed 25/03/2014.)

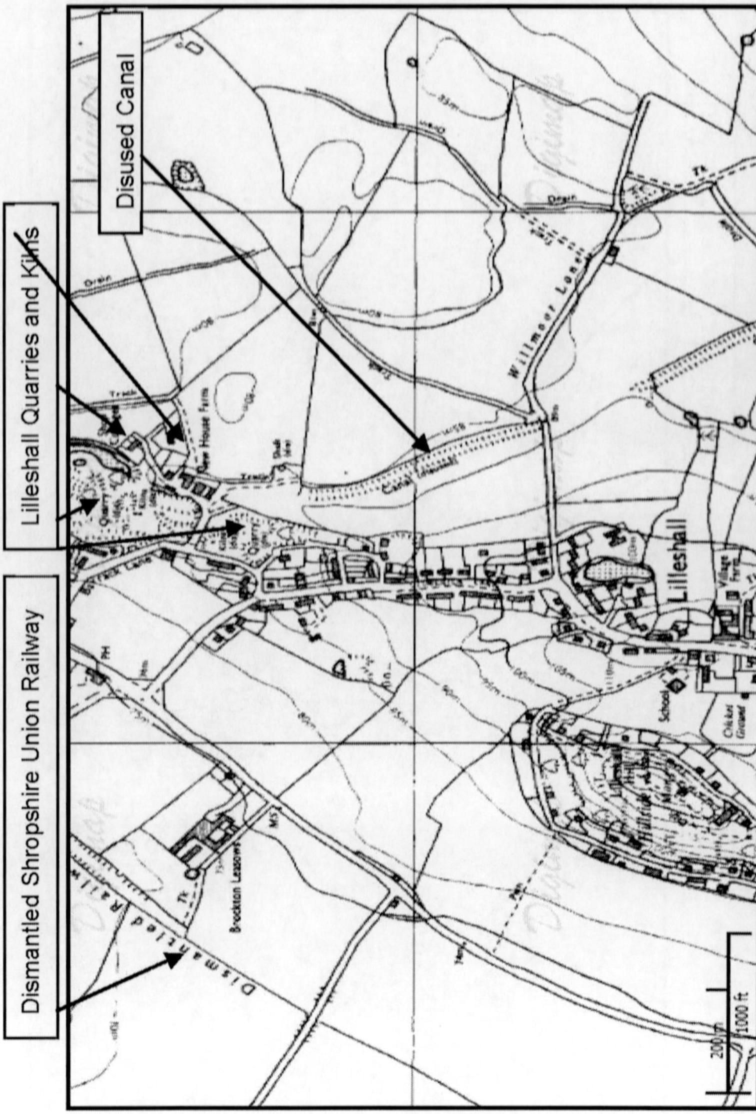

Figure 6.9: Lilleshall 1980. (Source: Digimap, www.ac.uk/ancientroam, accessed 25/03/2014).

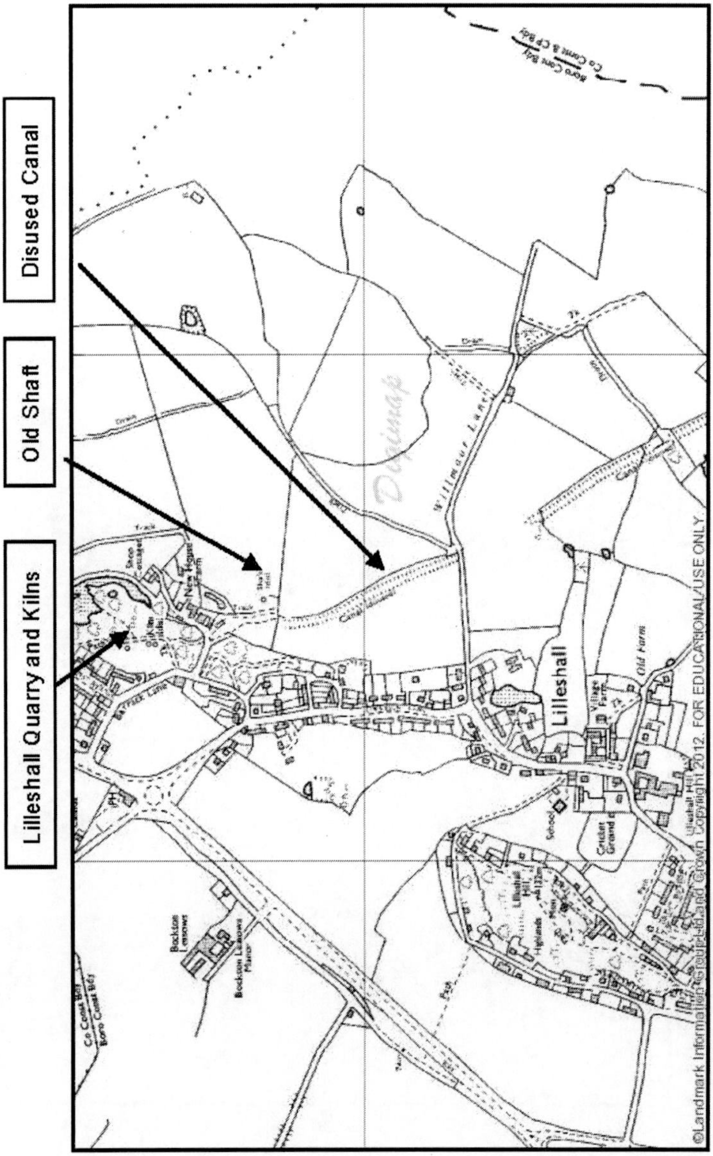

Figure 6.10: Lilleshall 1990. (Source: Digimap, www.ac.uk/ancientroam, accessed 25/03/2014.)

The 1890 map of Wenlock Edge area shows the first quarry near Roman Banks; the second quarry opposite the Methodist Church and a limekiln south of High Wood. The 1900 map of the same area show the same quarries with two additions one where the old limekilns were and the other near Under Heywood. The 1950 map of Wenlock Edge now only shows one quarry near Under Haywood. The 1970s map of the Wenlock Edge area now shows no quarries, however as the research indicates quarrying was taking place in the early part of the 21st century (Digimap, www.ac.uk/ancientroam, accessed 25/03/2014). Below are the maps of Wenlock Edge:

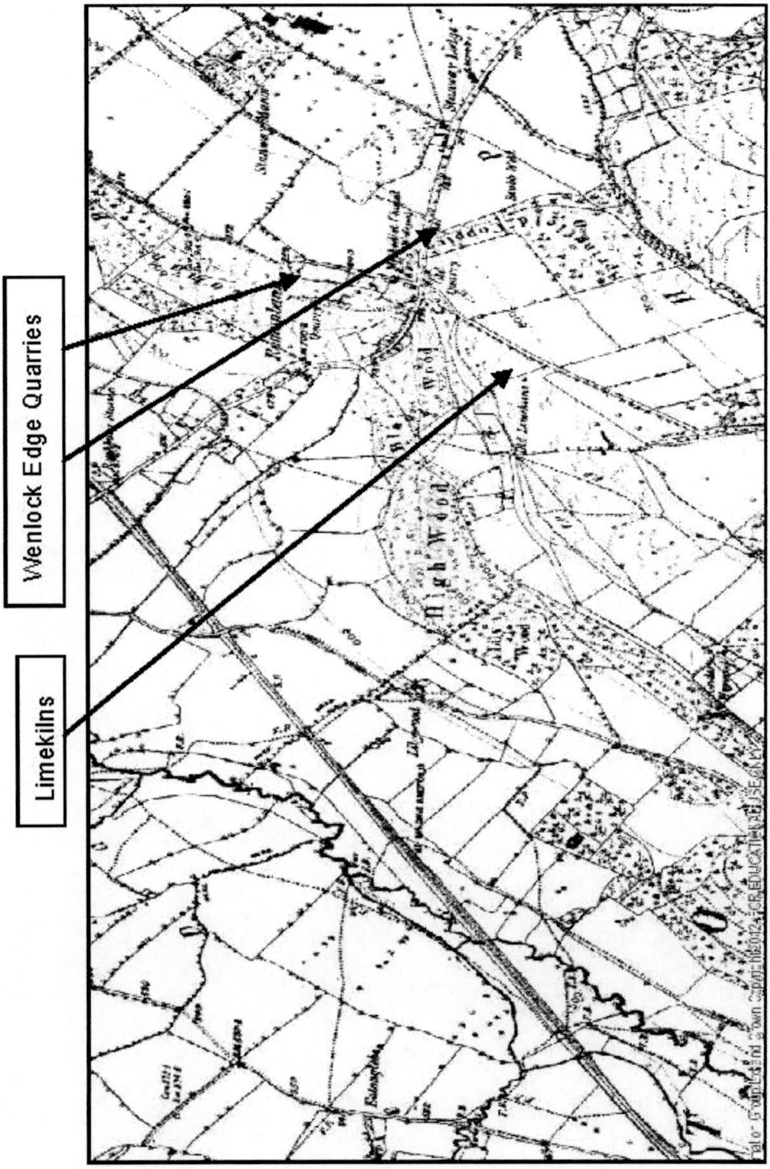

Figure 6.11: Wenlock Edge Quarries and Limekilns 1890. (Source: Digimap, www.ac.uk/ancientroam, accessed 19/03/2014).

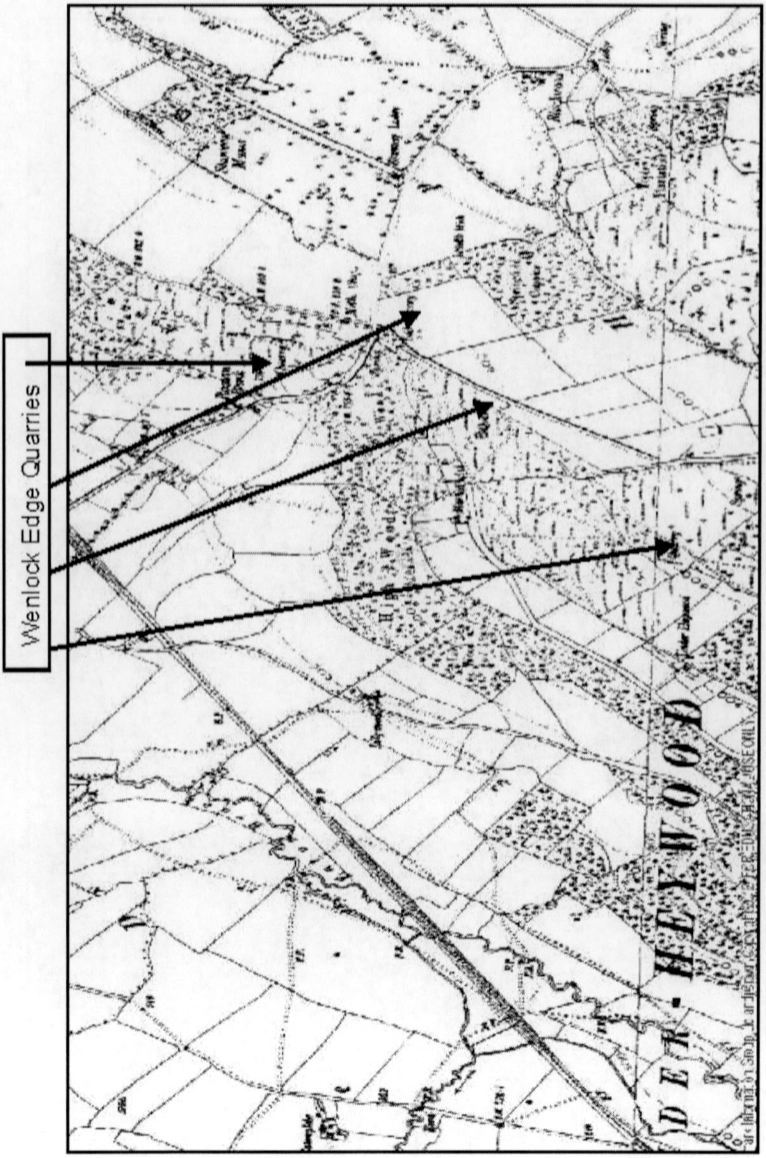

Figure 6.12: Wenlock Edge Quarries and Limekilns 1900. (Source: Digimap, www.ac.uk/ancientroam, accessed 19/03/2014.

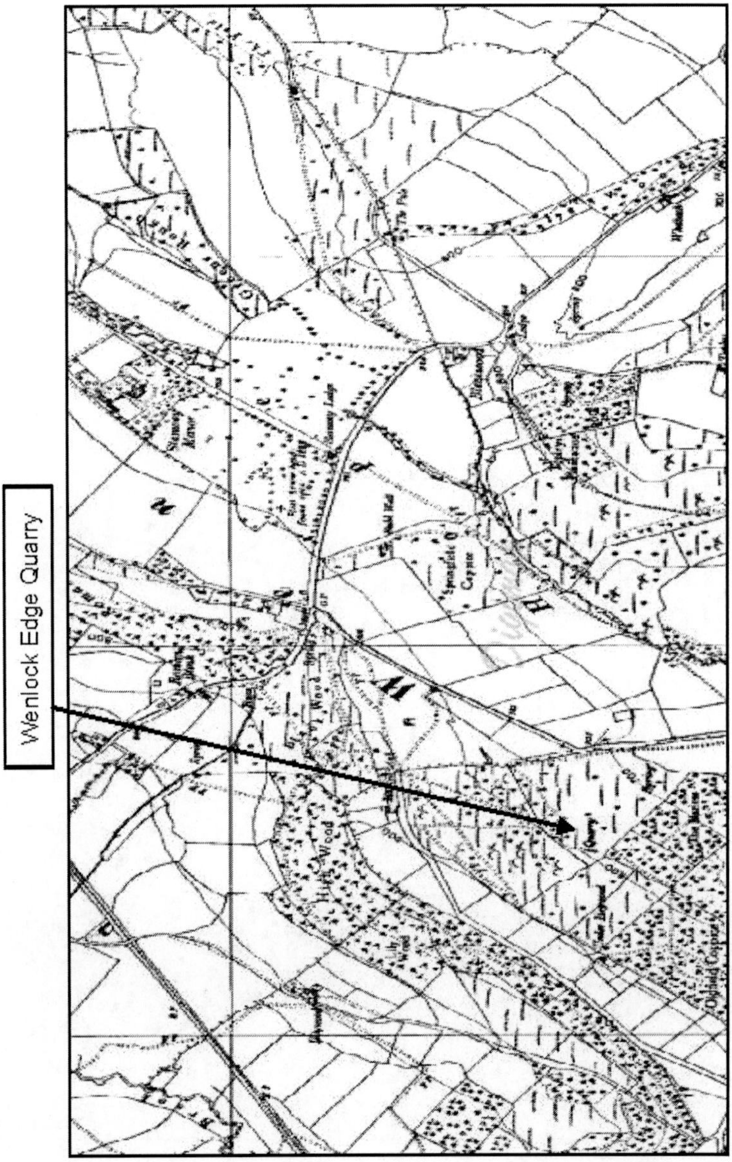

Figure 6.13: Wenlock Edge Quarries and Limekilns 1950. (Source: Digimap, www.ac.uk/ancientroam, accessed 19/03/2014).

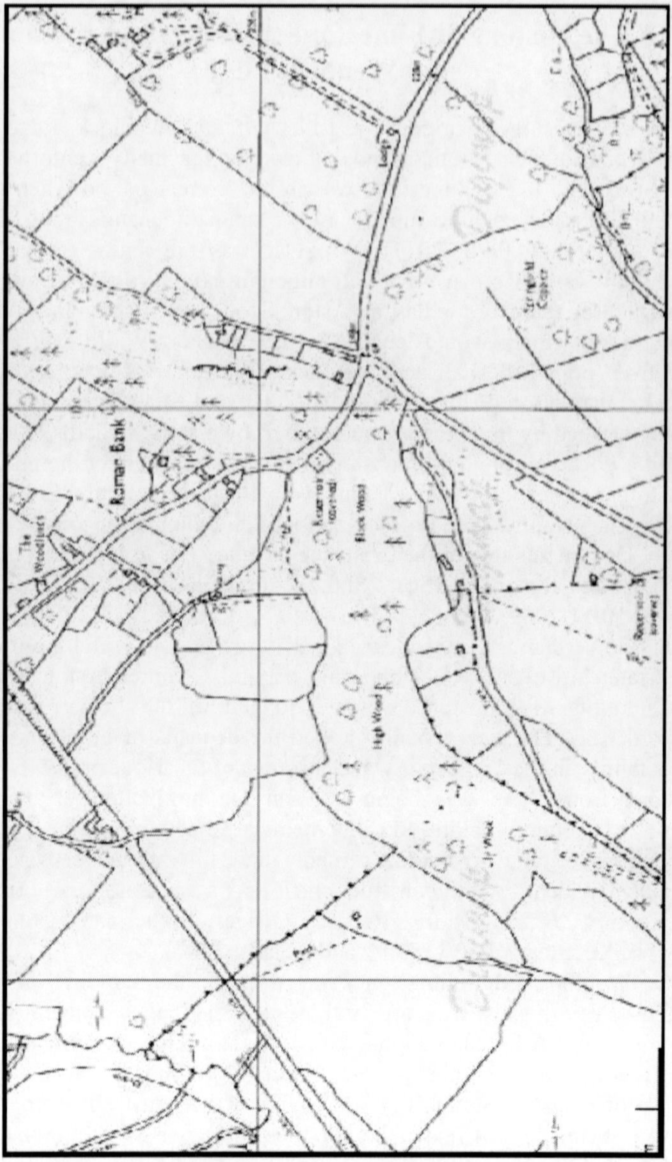

Figure 6.14: Wenlock Edge Quarries and Limekilns 1970. (Source: Digimap, www.ac.uk/ancientroam, accessed 19/03/2014.

6.6 The History of Limestone workings at Lilleshall and Wenlock Edge

Shropshire limestone workings at Lilleshall and Wenlock Edge have afforded employment to thousands of people for many centuries. The earliest recorded use of limestone was in the Roman period when it was quarried and used as lime mortar, rather than dimension stone (Roger White, per comm, 04/08/2011). Wroxeter was the major Roman city, Whitchurch was a Roman town and although extant villas in Shropshire are very rare, there are villas at Aston Scott, Lea Cross and Whitley Grange, all of which would have used lime plaster (White *et al*, 1998). Stone from both Lilleshall and Wenlock Edge quarries was used in the original construction of both Lilleshall Abbey and Much Wenlock Priory; possibly worked by the monks who cut their own stone (Adams, 2007 and Pinnell, 1999). Lilleshall Abbey, founded 1145-48, was the home of the Augustinian Order known as Canons, who lived there until 1538. Today the abbey is a ruin, yet the limestone blocks and the mortar joints can still be seen. Documents record the existence of lime kilns at Lilleshall in 1625 (Staffordshire Record Office, Sutherland Collection, D593/J/13/2/1, accessed 10/11/2009).

The Gower family, of Lilleshall had large deposits of limestone on their estate; however the neighbouring parish of Church Aston also had large amounts of limestone which belonged to the Leeke family of Longford Hall. The Leeke family leased the deposits of limestone to the Gower family in the 18th century and for most of the 19th century. In 1776, limestone from there was being used in the production of hydraulic cement, a ground-breaking development at that time. The chemical composition of the hydraulic cement made it set under-water and consequently ideal for the construction of docks and harbours during the 19th century (Staffordshire Record Office, Sutherland Collection, D593/1/1/32, accessed 10/11/2009 and Adams, 2007).

The Wenlock Edge limestone was used in the building of Much Wenlock Priory. This was originally established as a convent by St. Melburga about A.D 680, daughter of Merewald, King of Merica. In A.D 874 it was destroyed by the Danes, eventually being restored in 1017 by the Earl of Chester. Some 50 years later it was destroyed again by the Normans, being rebuilt under the benefice of Roger de Montgomery by monks of the Clunaic Order (Pinnell, 1999). In addition, the limestone was also used in the construction of Holy Trinity Church, Much Wenlock and Buildwas Abbey, in the 12th century. From 1399-1400 Wenlock Edge quarries also supplied stone for the construction of Caus Castle.

In the 16th and 17th centuries throughout Shropshire self-employed lime carriers were at work, particularly in the quarries at Wenlock Edge as part of the processes of lime burning which was carried out into the 18th century (Shropshire Archives, documents no, 3898/Rg/I, 1224/2/36 and 1224/2/116, accessed 23/04/2010). In the 18th century, illegal quarrying, together with the construction of unauthorized lime kilns, was in action on common land. During the middle of the 18th century until the early 20th century, the use of limestone intensified and limestone quarrying expanded at Wenlock Edge, becoming a major element in the economic growth of the local area and Shropshire. Over the period of 150 years, a combination of industrial and agricultural interests developed (Shropshire Archives, documents no, 3898/Rg/I, 1224/2/36 and 1224/2/116, accessed 23/04/2010). Both Lilleshall and Wenlock Edge quarries were now also providing limestone for a range of uses, not only for Shropshire, but for the rest of Britain (Scard, 1989).

As the iron industry expanded there was a greater need for limestone for smelting iron, so contributing to the growth of the quarrying industry at Lilleshall and Wenlock Edge. In the early 18th century Abraham Darby purchased limestone from Wenlock Edge for iron smelting in his furnaces; this continued with Abraham Darby IV and other Ironmasters throughout the 19th century (Currie, 1998, Williams, 1997 and Trinder 2016). By the end of the 19th century, Shropshire's iron industry was in decline, so the demand for Wenlock Edge limestone steadily diminished. This demise affected the limestone quarrying industry and consequently the quarry owners needed to renegotiate the leases during the early years of the 20th century, in order to reduce the cost of land rent, to save the industry. Even so, some quarries were either closed or mothballed (Currie, 1998, Williams, 1997 and Holmes, 1987). The other two products, limestone for agriculture and for construction, which have been around before the onset of the iron industry's rise and decline, were still prospering.

The First World War was a major factor in the demise of the quarrying industry; manpower, vital to the industry, was now needed at the front (Trinder, 2000). During the early to middle years of the 20th century, the limestone industry continued to turn over slowly, but towards the end of 1940s, after The Second World War, there was a need for limestone, which in turn revived the industry. In addition, new roads or improvement of existing ones were needed to help with the transportation of limestone across the country. Shropshire County Council, for environmental reasons, decided that quarrying should stop on Wenlock Edge, after each quarry had been worked out (Shropshire County Council, 1978. Structure Plan: Written Statement: Mineral Extraction on Wenlock Edge, accessed

14/09/2013). However in the 1980s the County Council accepted that, if replacement stone was needed and comparable stone could not be located elsewhere; then quarrying might have to be resumed on a small scale. In 2007 Lea Quarry, Wenlock Edge, was mothballed and closed, yet all the machinery associated with a modern working quarry still remains on site (The Quarry Managers' Journal, 1968, Williams, 1997 and Mineral Extraction on Wenlock Edge, 1978).

6.7 Lilleshall Estate and Wenlock Edge Owners

The potential to extract limestone from monastic lands was realised after the dissolution of the monasteries when prominent entrepreneurs purchased these lands. This happened at Wenlock Priory in 1540 when purchase was made by Lawrence Benthall, Sir Basil Brookes, James Clifford and John Weld, and at Lilleshall Abbey in 1538, two years earlier, by James Leveson a single entrepreneur (Alfrey *et al,* 1993, Kelly's Directory, 1985 and Yates, 2008). Over time the lime industry expanded into different industrial areas.

The Leveson family, who had gained their fortune in the Wolverhampton wool trade during the 15th and 16th centuries, purchased the Lilleshall Abbey lands following the Dissolution of the Monasteries in 1538 (Yates, 2008). A deed dated 24th December 1539, shows that Lilleshall Abbey and its lands were sold to James Leveson; initial payment was made by way of a lump sum, followed by annual instalments. Similarly, one year later, in 1540, Wenlock Priory was dissolved and the priory lands were purchased from the crown by private landowners, Lawrence Benthall, Sir Basil Brookes, James Clifford and John Weld, who were aware of the mineral potential under the former priory lands (Alfrey, *et al,* 1993).

The Leveson family of Lilleshall set to work draining the land at Weald Moor and creating a causeway across the marshy ground, so allowing for general agricultural improvements and the development of limestone extraction. However, it was not until the 18th century that the Lilleshall Limestone industry started to be of significance. Documents held at the Ironbridge Gorge Museum Trust relate to the quarrying and mining activities of the Leveson family and the Lilleshall Company. These documents reveal the extent of financial and contractual arrangements associated with the leasing of quarry sites, industrial buildings, accommodation and tonnage of the minerals, particularly white and grey limestone (Ironbridge Gorge Museum Trust Collection of Documents

Relating to The Lilleshall Iron & Steel Company Ltd 1660-1990. Ref: 1998.320(DLIL/3) 154.29 Sept 1797).

In 1746, Sir John Leveson's son, also known as John, became Earl Gower. It was not until his son, Granville Leveson-Gower, became the second Earl in 1754 that the industrial story began. In his time he became a major landowner having estates at Lilleshall, Sherrifhales, Donnington Wood, St. George's, Priorslee, Wombridge and Snedshill. At the same time he was a Member of Parliament, Lord of the Admiralty, Lord Privy Seal and Lord Chamberlain to King George III. Gower's second wife was Lady Louisa Egerton, sister of Francis, 3rd Duke of Bridgewater, who inherited the title when still a minor. He became the ward of the Earl Gower and the Duke of Bedford (Yates, 2008 and Adams, 2007). Francis, 3rd Duke of Bridgwater, was the pioneer of canal building, thus bringing about radical changes to the transport system, which in turn benefited the industrial revolution. In 1757 John Gilbert, the Duke of Bridgewater's engineer, was responsible for introducing his brother Thomas, a barrister and Member of Parliament, and legal adviser to Earl Gower. John Gilbert, as steward to the 3rd Duke of Bridgewater, came into contact with engineer James Brindley, who worked on the construction of the Bridgewater canal; thus gaining valuable experience in canal construction. This experience would in time work to the benefit of Earl Gower (Shropshire Archives, Lilleshall Records, accessed 07/08/2009).

In 1764 Earl Gower formed a partnership with the Gilbert brothers, known as the Lilleshall Partnership, forming The Earl Gower and Company, which was their trading name until 1786. The partnership would last for 34 years until the death of John Gilbert in 1795 and Thomas Gilbert in 1798. Lord Gower had 50% of the shares in the company, as he had provided the capital and land. The other 50% of the shares belonged equally to the Gilbert brothers, who provided the knowledge for canal planning, buildings and legal advice. The partnership constructed railways, navigable canals and sluices; in particular the Donnington Wood Canal which was started in 1765 and completed in1767 by John Gilbert engineer transporting limestone and coal to fire the lime kilns at the Lilleshall estate (Staffordshire Record Office, Sutherland Collection, D593/B/2/7/22/1, D593/B/2/7/22/2, D593/J/13/2/1, accessed 10/11/2009 and I.B.G.M. L, Ove Arup *et al,* 1987). The records show that in December 1805, Lilleshall quarries and mines were producing 3,700 tons of limestone. Lilleshall and Church Aston limestone works were a combination of quarries and mines; the deepest shaft being 75.59 metres in extent (Staffordshire Record Office, Sutherland Collection, D593/J/13/2/1, accessed 10/11/2009).

6.7.1 Different Industries Associated with Wenlock Edge Limestone

Different industries were associated with the limestone industry in the Much Wenlock area involving large or small quarry owners, independent lime burners, and mineral merchants. From Historical Directories, Shropshire Archive and Kelly's Directory, dating from 1851 to 1934, lime burners are recorded as working in Much Wenlock Parish. Mr Jeremiah Cooper was a lime burner from 1851 to 1895, a period of forty-four years at a place known as The Bank, Wenlock. Then there are a number of people recorded as lime burners from 1859: Elizabeth Edwards, Edward Harrington, Frances Harrington, Thomas Harrington and Thomas Jukes of Wenlock Edge. William Jukes, Edward Mason, John Nicklin, John Onions and George Sheppard of Wenlock Edge.

It is interesting to note that two women are recorded as being lime burners, Elizabeth Edwards and Frances Harrington. The second woman has the same surname as Edward Harrington and Thomas Harrington; they could be a family working together as lime burners on Wenlock Edge. In 1859 Thomas Harrington was noted as being a lime burner and again, nine years later in 1868, is recorded as being a lime burner; however, this time no Edward Harrington or Frances Harrington appear on the register. It is possible that they were Thomas Harrington's, father and mother. Mr Thomas and William Jukes in 1859 are recorded has being lime burners, at Wenlock Edge, and at Coates, Wenlock Edge. However, Thomas Jukes is also recorded as farming Coates Farm in 1851; as mentioned before they could be brothers. From my research, a tithe map of 1847 indicates a small disused quarry near the summit of Wenlock Edge, north-west of Coates Quarry. I suggest that these two brothers were working together on Coates Farm and lime burning at Wenlock Edge. In 1868, another two people: Edwin Hayward of Priesthope, Much Wenlock, and Nathaniel Reece of Easthope, Much Wenlock, are also recorded as being lime burners (Historical Directories. www.historicaldirectories.org, accessed 03/06/2010 and Shropshire Archives, Kelly's Directory of Shropshire, accessed 11/06/2010).

In the year 1880, Mrs Letitia Hayward was recorded as being a lime burner at Presthope, Mush Wenlock. My research shows that Mrs Letitia Hayward, 74 year old, was the lessee of Knowles Quarry and is recorded as a widow, presumably, of Edwin Hayward in that same year. Presthope and Easthope were found on the B4371, where Knowles Quarry was located. Knowles Quarry also marks the starting point for a circular walk, which starts out from the National Trust car park, south-west of Much

Wenlock. This is where Mrs Letitia Hayward in 1880 was a lime burner, possibly at the same location as Knowles Quarry. In 1863, Mr Thomas Jukes took control of Coates Quarry and is recorded in 1880 at Coates, Stone Quarries, Presthope, Much Wenlock. Also, in 1880 Joseph Norry is recorded as working as a lime burner at Black Lane, Much Wenlock (Historical Directories. www.historicaldirectories.org, accessed 03/06/2010, Shropshire Archives and Kelly's Directory of Shropshire, accessed 11/06/2010).

In 1891, Henry Hayward, George Hayward and Elizabeth Hayward are shown as lime burners at Presthope, Much Wenlock. Henry Hayward and George Hayward were in all probability the sons of Letitia Hayward as from my research we know that she employed them at Knowles Quarry, with her other son George, grandson Edward and granddaughter Mary Cottrel. Elizabeth Hayward could either be a daughter, granddaughter or wife of one of her sons. In 1891, Thomas Edward Jukes is shown working as a lime burner; it could be conjectured as to whether he might be the son of Thomas Jukes of 1859, and so possibly carrying on the family business (1881 England Census, www.search.ancestry.co.uk, accessed 14/09/2010 and Williams, 1997).

In 1891, John Bolt's occupation is lime burner; he is also recorded as manager for Lilleshall Company Ltd at Presthope, Much Wenlock, Lilleshall and Newport. My research of the Lilleshall records for 1862, shows that Earl Granville Leveson-Gower of Granville leased 28 acres of land between Hilltop and Presthope, at Wenlock Edge, from Moses George Benson of Lutwyche Hall. There is no evidence to say if he was manager of the Presthope Quarry, Much Wenlock, however it is a possibility (Shropshire Archives, Lilleshall Records, 809/6/1, 809/6/2 809/6/4 and 809/6/5, accessed 05/08/2009). In 1891, Lloyd G. Westwood of Much Wenlock was recorded as a lime burner, as was Mrs Eliza Nicklin of Stretton Road, Much Wenlock. In 1859 John Nicklin was working as a lime burner, between the two dates there is thirty-two years difference; it is possible that they might be related. In 1891, both Mardol, J. Norry and Richard Onions are recorded as working as lime burners at Much Wenlock; it is possible that the latter may have been related to the earlier 1859 John Onions, lime burner. In 1891, Mr William Hales was working as a Lime Merchant at Barrow Street, Much Wenlock. In 1895 Mr William Hales is again recorded, however this time he seems to have expanded his business into Lime, Coal and Salt Merchant and Lime burner. Between the years 1891 to 1895 Mr William Hales had expanded his business from just a Lime Merchant by selling other minerals. In the late 19th century, the Lime, Coal and Salt that he sold would have been in great demand from people in lime-burning occupations. Lime burning was a very dangerous job. As well as being used for its heating properties in the home, coal was used as fuel in lime burning. At both dates Mr William Hales is recorded as being at Barrow Street, then Barrow Street Works, The Knowle and Presthope, Much Wenlock. In 1891, The Lilleshall Company Ltd were also Lime Merchants in Presthope, Much Wenlock. Having two Lime Merchants in Much Wenlock would have brought them both into competition with each other, especially with Lilleshall Company Ltd being a considerably larger business with their own limestone quarries (Historical Directories. www.historicaldirectories.org, accessed 03/06/2010, Shropshire Archives and Kelly's Directory of Shropshire, accessed 11/06/2010).

Over a 39 year period, the years 1895, 1917, 1922, 1929 and 1934, record Boulton A. and Company A. Boulton and Company as having a business as Lime Burners and Coal Merchants of Farley, Much Wenlock. Farley is on the A4169, on the north side of Much Wenlock, which leads to Ironbridge whereas the B4371 is on the Much Wenlock to Church Stretton Road, where the quarries were located. Farley is, however, conveniently situated for the Shrewsbury to Bridgnorth railway. In 1913, Harry J. Wild is recorded as being the District Manager of The Lilleshall Company Ltd, which refers to the company as being quarry owners and lime burners at Presthope, Much Wenlock, where we know that they operated quarries. Additionally in 1913, G. Meakin, Shaw and Kendrick, are also recorded as Lime Burners of Stretton Road, Much Wenlock. Starting in 1895, William Hayes took over the lease of Knowles Quarry. Five years later in 1900, William Hayes and Son were the proprietors of Knowles Quarry. The research shows that William Hayes was expanding from just a Lime Merchant into quarrying, so that he was then able to both quarry and sell his own limestone. The years, 1900, 1913, 1917, 1922 and 1926 record William Hayes and Son as Quarry Owners, Lime, Coal and Salt Merchants, and also Lime Burners of Barrow Street, The Knowle and Presthope, Much Wenlock. Following this in 1934 and 1937 the company was then known as William. M. Hayes and Son, still Quarry Owners and Lime Burners until eventually in 1937, William. M. Hayes and Son became a limited company. Could it be that William M. Hayes was the son of William Hayes of 1891 who had then taken over the business? The book goes into more detail regarding the Hayes family and their connection with the quarrying industry. Finally in 1964, Hayes' Quarry changed hands being purchased by the Ridge Limestone Ltd (Historical Directories. www.historicaldirectories.org, accessed 03/06/2010, Shropshire Archives and Kelly's Directory of Shropshire, accessed 11/06/2010).

In 1922, 1926, 1929 and 1934, Moore and Shield, operated from 1937 to 1941 as Limestone Quarry Owners, Road Stone and Chipping Speciality, Stone for Blasting Furnaces and Lime Burners, of Westwood Quarries, Much Wenlock. At some time in the mid-1920's, Kane purchased the Knowles Lime Works, Much Wenlock, and in 1934 and 1937, C.R.H. Kane is recorded as being the Limestone Quarry Owner of this same lime works. My book contains more information regarding Kane in the quarrying industry. However, in 1922, Bonum Fertilisers Company was listed as Limestone Quarry Owners, Lime Burners and Fertiliser Merchants of Knowles, Much Wenlock. It is not known whether this is Knowles Quarry, because no other reference to them can be found. There is a possibility that there was some other quarry of the same name, their main business could have been fertiliser selling and buying because there would have been plenty of opportunity for buying processed lime from other quarry companies. The last person recorded is Edwards Sankey who, in 1926, 1929 and 1934 was Limestone Quarry Owner at Coates Lime Works, Much Wenlock. In my book Mr Edwin Sankely is recorded working at Coates Quarry in 1914; who paid £10 per year to George Lloyd until his death in 1924. I suggest that Edward Sankey and Mr Edwin Sankely are the same person, having been recorded with just a different spelling in different records. This table provides a unique insight and cross references to my research on quarrying in Much Wenlock. It shows how the quarrying industry and associated industries grew in the 19th and 20th centuries in Much Wenlock, Shropshire. Much Wenlock was a hive of quarrying and lime burning activities; from the table one can see how it grew in particular for example the Hayes family (Historical Directories, www.historicaldirectories.org, accessed 03/06/2010, Shropshire Archives and Kelly's Directory of Shropshire, accessed 11/06/2010). Compiled below is a list of individuals and groups of people that benefited from the limestone industry in the Much Wenlock area:

Table 6.2: List of Lime Burners, Merchants and Quarry Owners of The Much Wenlock Parish:

Names:	**Occupations:**	**Places:**
Cooper, Jeremiah.	Lime Burner.	1851 and 1895, The Bank, Wenlock.
Edwards, Elizabeth.	Lime Burner.	1859, Wenlock.
Harrington, Edward.	Lime Burner.	1859, Wenlock Edge.
Harrington, Frances.	Lime Burner.	1859, Wenlock Edge.
Harrington, Thomas.	Lime Burner.	1859, Wenlock Edge.

Jukes, Thomas.	Lime Burner.	1859, Wenlock Edge.
Jukes, William.	Lime Burner.	1859, Coates, Wenlock Edge.
Mason, Edward.	Lime Burner.	1859, Bull Ring, Wenlock.
Nicklin, John.	Lime Burner.	1859, Wenlock Edge.
Onions, John.	Lime Burner.	1859, Wenlock Edge.
Sheppard, George,	Lime Burner.	1859, Wenlock Edge.
Harrington, Thomas.	Lime Burner.	1868, The Bank, Wenlock.
Hayward, Edwin.	Lime Burner.	1868, Priesthope, Much Wenlock.
Reece, Nathaniel.	Lime Burner.	1868, Easthope, Much Wenlock.
Hayward, Letitia.	Lime Burner.	1880, Presthope, Much Wenlock.
Jukes, Thomas.	Coates, Stone Quarry.	1880, Presthope, Much Wenlock.
Norry, Joseph.	Lime Burner.	1880, Back Lane, Muck Wenlock.
Hayward, Henry, Hayward, George and Hayward Elizabeth,	Lime Burners.	1891, Presthope, Much Wenlock.
Jukes, Thomas, Edward.	Lime Burner.	1891, Stretton Road, Much Wenlock.
Bott, John (Manager). Lilleshall Co. Ltd.	Lime Burners.	1891, Presthope, Much Wenlock, Lilleshall and Newport.
Westwood , Lloyd, G.	Lime Burner.	1891, Much Wenlock.
Nicklin, Eliza, Mrs.	Lime Burner.	1891, Stretton Road, Much Wenlock.
Norry, J. Mardol,	Lime Burner.	1891, Much Wenlock.
Onions, Richard.	Lime Burner.	1891, Wood House Farm, Much Wenlock.
Hayes, William.	Lime Merchants.	1891, Barrow Street, Much Wenlock.
Lilleshall Co. Ltd.	Lime Merchants.	1891, Presthope, Much Wenlock.
Hayes, William.	Lime, Coal and Salt Merchants; Lime Burners.	1895, Barrow Street, works, The Knowle and Presthope, Much Wenlock.
Boulton, A. and Co.	Lime Burners and Coal Merchants.	1913, Farley, Much Wenlock.
Wild, J, Harry. (District Manager). Lilleshall Co. Ltd.	Quarry Owners and Lime Burners.	1913, Presthope, Much Wenlock.

Meakin, G.	Lime Burner.	1913, Stretton Road, Much Wenlock.
Shaw and Kendrick.	Lime Burners.	1913, Stretton Road, Much Wenlock.
Boulton, A and Co.	Lime Burners.	1895, 1917, 1922, 1929, 1934, Farley, Much Wenlock.
Hayes, William and Son.	Quarry Owners. Lime, Coal and Salt Merchants. Lime Burners.	1900, 1913, 1917, 1922 and 1926 Barrow Street, The Knowle and Presthope, Much Wenlock.
Moore and Shields. Moore and Shield (Kane, C.R. Proprietor).	Limestone Quarry Owners, Road Stone and Chippings, Speciality, Stone for Blast Furnaces. Lime Burners.	1922, 1926, 1929, 1934, 1937 and 1941 Westwood Quarries, Much Wenlock.
Bonum Fertilisers Co.	Limestone Quarry Owners. Lime Burners. Fertiliser Merchants.	1922, Knowles, Much Wenlock.
Sankey, Edwards.	Limestone Quarry Owner.	1926, 1929, and 1934, Coates Lime Works, Much Wenlock.
Hayes, W. M. and Son. Hayes, W. M. and Son Ltd.	Limestone Quarry Owners. Ditto and Lime Burners.	1934, Presthope, Much Wenlock. 1937, Presthope, Much Wenlock.
Kane, C. R. H.	Limestone Quarry Owner.	1934 and 1937, Knowles Lime Works, Much Wenlock.

(Source: Historical Directories. www.historicaldirectories.org, accessed 03/06/2010 and Shropshire Archives, Kelly's Directory of Shropshire, 1913, 1917, 1922, 1926, 1929, 1934, 1937 and 1941, accessed 11/06/2010).

6.8 Industrial Growth of Lilleshall and Wenlock Edge

During the 16th and 17th centuries, after the social upheaval of the Reformation and the English Civil War, a sustained period of economic growth and development took place throughout the 18th century and into the 19th century. Those who had the wherewithal to exploit these new industrial opportunities did so; particularly the iron masters who required more limestone, which was used as flux in the blast furnaces, resulting in the eventual expansion of the iron industry. Consequently new resources of limestone were required to supply this demand, which in turn contributed to the development of the limestone industries at Lilleshall and Wenlock Edge and the advancement of new technologies. The middle of the 19th century to the early 20th century continued to witness substantial growth and technological developments such as power cranes, stone dressing and crushing machineries, which advanced the methods of quarrying (Shropshire Archives, Lilleshall Records, accessed 07/08/2009 and Alfrey, *et al,* 1993 and Stanier, 2000). Therefore individuals that once owned and operated small quarries and kilns were swallowed up by large landowners and their companies, as happened at Lilleshall and Wenlock Edge. The links between landowners and the mineral industries gave many large estate owners such as Earl Gower, the Duke of Bridgewater and Lord Dudley the opportunity of increasing their rents, profits and royalties (Musson, 1978 and Mingay, 1979).

In 1709 Abraham Darby of Coalbrookdale built the first blast furnace fuelled by coke instead of charcoal. The 17th century saw the iron industry of Shropshire expanding so limestone was in greater demand, and the quarrying industry at Wenlock Edge began to develop. However, by the middle of the 17th century there was a growing acceptance of coke-fuelled furnaces within the iron industry. The use of charcoal enabled the blast furnaces to produce higher temperatures, which increased the iron production. By the end of the 17th century the amount of pig iron produced was nearly 90,000 tons per year, compared with approximately 28,000 tons per year in the middle of the 17th century. By the 1830s, production levels in the iron industry were 700,000 tons per year. Consequently this development required the quarrying industry to produce more limestone for fluxing, which in turn meant additional quarries and a larger workforce; this resulted in the expansion and development of both industries. Wenlock Edge and Bentall Quarries were perfectly located for this purpose (More, 1997).

The height of production in the limestone quarrying industry occurred in the 19th century, but following the decline of the local iron industry there was an equal decline in limestone quarrying. By the early 20th century, Ironbridge limestone quarries were exhausted, but Wenlock Edge limestone quarries remained very productive, which continued into the later part of the 20th century (William, 1997, Holmes, 1987 and Mineral Extraction on Wenlock Edge, 1979).

6.9 Lilleshall and Wenlock Edge Limestone Workings

In 1799 a contract was drawn up between Granville Leveson-Gower 1st Marquis of Stafford and Mr Thomas Botfield to supply 9,000 tons of limestone per year for the next 30 years. The price of the limestone was 4 shillings and 6 pence a ton, which was to be transported by way of the Donnington Wood Canal. The contract indicates how developed the limestone mines were, to contemplate taking such an order for this period in time (Gale *et al,* 1979 and The Wellington Journal, 1860). Below is a map of Lilleshall with its surrounding areas showing the location of the Lilleshall Company quarries and mines:

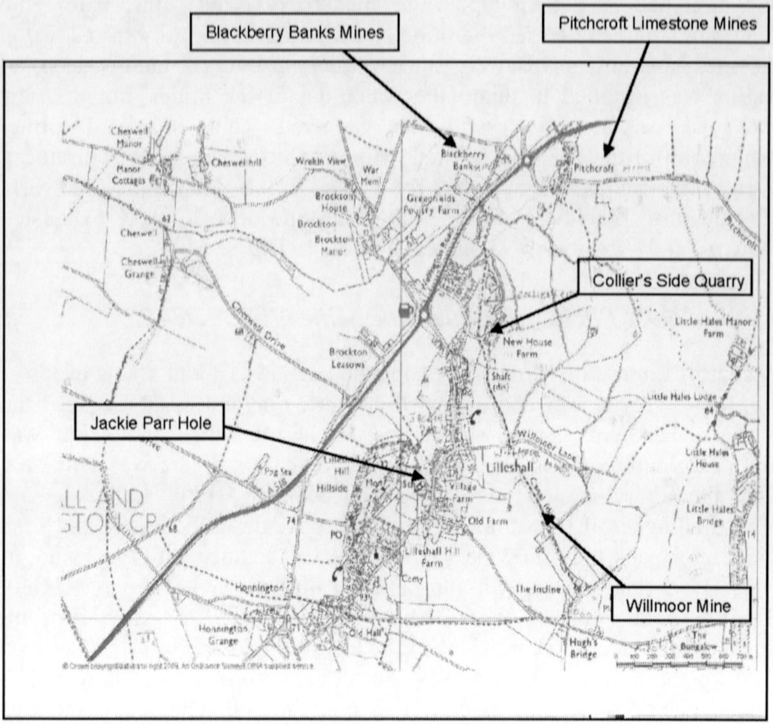

Figure 6.15: Lilleshall map showing location of quarries and mines (Source: University of Birmingham Map Library accessed 09/05/2010).

6.9.1 Blackberry Bank Mines NGR: SJ739172

At the end of the 17th century, limestone quarrying and mining was well established in the Lilleshall area. The Blackberry Bank Mine limestone deposits belonged to the Leeke family of Longford Hall in Church Aston, which was subsequently leased to Lord Gower's Lilleshall Company, for most of the 18th century and early 19th century. The Blackberry Bank Mine was connected by way of a canal branch constructed to the east of the village linking to the Lilleshall Company's main canal system (Staffordshire Record Office, documents no, D593/1/132 and D593/M/17/1, accessed 10/11/2009). In the early 18th century mining operation had been taking place at Blackberry Bank Mine, before the Lilleshall Company leased the mine and tried unsuccessfully to redevelop the site. Two new shafts were sunk; one being over 37 metres deep, an engine was installed to pump the water out of the mines, but the new enterprise was less successful than expected. Thus ended all further exploration of the Blackberry Bank Mines, which had probably produced over a million tons of limestone? By the end of the 18th century and early 19th century, Blackberry Bank Mine limestone deposits were exhausted (Adams, 2007, Pearce, 1995 and Gale, et al, 1979).

6.9.2 Pitchcroft Limestone Mine NGR: SJ741173

Pitchcroft Limestone Mine was established in 1798; it had seams of lower red, upper grey, lower and upper white limestone and located east of the former Blackberry Bank Mine. The Pitchcroft Limestone Mine was producing approximately 20,000 tons per year. This figure was estimated from the Church Aston and Lilleshall accounts of 1798. Calculated on what the Pitchcroft Limestone Mine was producing in the last years of the 18th century and the following years of the 19th century until its closure in 1860 (Ironbridge Gorge Museum Trust, Lilleshall Collection, 1998.320, (DLIL/3) 556a, accessed 05/08/2009, Ove Arup *et al,* 1987 and Shropshire Site and Monument Records No: 7158, accessed 07/08/2009).

Two shafts were sunk at the Pitchcroft Limestone Mine. One was 137 metres deep, connecting to extensive underground galleries. Limekilns were built near to the entrances of the shafts. Later, under a special agreement with Earl Leveson Gower of Granville the Lilleshall Company were allowed to sink one further shaft on land owned by the Leeke family and under the tenancy of Mr William Jackson. The limestone works at Pitchcroft were connected to the Donnington Wood Canal by the Lilleshall canal branch, via Willmore Bridge to the canal wharfage near the

Donnington Furnaces. In 1846, the transporting of limestone was made easier when the Stafford to Wellington railway was built, the line passed near to the Pitchcroft Limestone Mine main shaft. In the 19th century accidents in all mines, coal and limestone were frequent; one too many accidents happened at Pitchcroft Limestone Mine, one in particular in 1858 when three men were killed after a giant layer of limestone collapsed on them (Ironbridge Gorge Museum Trust, Lilleshall Collection, 1998.320, (DLIL/3) 556a, accessed 05/08/2009). Although The Pitchcroft Limestone Mine was very productive throughout its life, in 1860 old workings started to flood, making it too dangerous to continue working. Following the closure of the mine, the Lilleshall canal branch that once connected the Pitchcroft Limestone Mines of became neglected and in time was blocked of at Willmoor Bridge. A map of 1882, the area shows the existence of the former Pitchcroft Limestone Mine. Today, the Pitchcroft Limestone Mine shafts and parts of the limekilns are still visible; the shafts are now covered with a metal grid; one is brick lined and the other stone lined (Robert Galloway, Lilleshall Site Visit, 22/08/2009, Shropshire Site and Monument Records No: 7158, accessed 07/08/2009 and Adams, 2007).

6.9.3 Colliers Side Quarry NGR: SJ735165

Colliers Side Quarry consists of two quarries, the larger one to the northeast and the smaller to the west side of Limekiln Lane; it was considered to be the Lilleshall Company's major quarry. The first evidence of a quarry at Colliers Side was in the late 18th century, a small shallow opencast bed, which eventually went underground. January 1805, Colliers Side Quarry produced 7,400 tons of limestone. Some seventy-five years later, 1877-1882 records show that Colliers Side Quarry was still producing limestone, but production is declining (Adams, 2007 and Shropshire Site and Monument Records No: n/n, accessed 07/08/2009). Table 6.3 shows the number of men working underground and above ground at Colliers Side Quarry from the years 1877 to 1882. In 1881, eighteen men were working both underground and above ground; this is the only time the number of men was equal. The limestone they extracted in 1881 was 7,184 tonnes. In 1877 there were nineteen men working underground and seventeen men working above ground, they mined 8,274 tonnes of limestone. In 1878 the number of men working underground dropped to eighteen as did the number of men working above ground to sixteen men, this was one man less for both columns. They mined 7,716 tonnes of limestone in 1878 a decrease of 558 tonnes per year from 1877.

In the year 1879 there were eighteen men working underground, this was the same as the year before; the number working above ground decreased to sixteen again; this was the same as the year before. They mined 7,368 tonnes of limestone; a fall yet again of 348 tonnes from the previous year. The year 1880 shows that there were eighteen men working underground the same as the previous year, and the number of men working above ground had increased by one to seventeen from 1878. The limestone they mined was 7,272 tonnes, a decrease of 96 tonnes. In the last year, 1882, the number of men working underground fell to sixteen, a decrease of two men from 1882, and three from 1877; men working above ground also fell by four in 1881, and three in 1877. The limestone mined was not recorded for the year 1882; however if it followed the previous year's decrease in quantity of the limestone mined, then again it would have fallen. From the year 1877 the tonnage of limestone mined and processed was 8,274 tonnes and in 1881 it was 7,184 tonnes a decrease of 1,090 tonnes of limestone. As we can see from the table 6.3 the number of men working underground stayed the same, apart from 1881 when it fell by one man. However, the number of men working above ground decreased every year apart from in 1881.

As the records of Lilleshall show, mining for limestone was a dangerous job for the men who worked underground. At Pitchcroft Limestone Mine there were many accidents, particularly in 1858 when three men were killed by a giant layer of limestone which fell on them. There was also the constant risk of flooding at this same mine. As this had happened at Pitchcroft, then there was every possibility that it might happen at Colliers Side Quarry and Mines.

The reason for the decrease in the number of men working above ground may have been the use of more efficient machinery, making it easier to process the limestone. It would be informative to know what pay the men working underground received, as compared to that received by the men working above ground. An analysis of these factors would show how the pay reflected the dangers of working underground in the mines. This table gives a unique record of the working life of men, both working under and above ground at Colliers Side Quarry and Mine (Ironbridge Gorge Museum Trust, accessed 05/08/2009 and Shropshire Archives, accessed 10/11/2009). A table below shows how many men were working underground and above ground, plus what they were producing in tonnage:

Table 6.3: Colliers Side Quarry and Mine

Year	No. of Men Under Ground	No of Men Above Ground	Tonnage per Year
1877	19	17	8,274
1878	18	16	7,716
1879	18	16	7,368
1880	18	17	7,272
1881	18	18	7,184
1882	16	14	No record

(Source: Ironbridge Gorge Museum Trust, accessed 05/08/2009 and Shropshire Archives, accessed 10/11/2009).

The Colliers Side Quarry and Mine underground workings were linked by inclines to tramways which connected to the surface. Extracted waste material was used to form immense embankments around the quarry. Constructed across the top of the embankments, into which the kilns had been built, were tramways that fed the kilns through two access tunnels. The tramways also carried processed material to Donnington Wood Canal. The Colliers Side Quarry Mine could also be accessed through three other shafts that had been sunk in other areas of the quarry ranging from 45 metres to 75 metres deep, in order to access the lower limestone deposits. In addition, a trench was excavated which continued underground in the area of Limekiln Lane (Robert Galloway, Lilleshall Site Visit, 22/08/2009, Shropshire Site and Monument Records No: n/n, accessed07/08/2009 and Adams, 2007).

White limestone was quarried extensively from the east of the woodland, in Lilleshall. The limestone extracted from the Colliers Side Quarry was of the highest quality and therefore primarily used for fluxing in the expanding iron-smelting industry. Around the former Colliers Side Quarry site there have been a number of subsidences, the most recent in the 1980's, to the east of the quarry, which was surveyed and backfilled (Ironbridge Gorge Museum Trust, Ove Arup *et al,* 1987).

In 1810, near the Colliers Side Quarry, the Lilleshall Company built 12 dwellings for their quarry workers, known as the Barrack Houses. Each house was 7.62 metres by 3.66 metres and one storey high, comprised of a living room, bedroom, store room. Outside there were three common washrooms, located at each end and one in the middle of the row of houses. The last occupant left in 1947; the houses remained empty for twenty-one years before being demolished in 1968. Below is a picture of the Barrack Houses:

Figure 6.16: Lilleshall Company Barrack Houses. (Source: Brown, 1976).

Other buildings on the Colliers Side Quarry site were offices, boat workshop, stables and weighing house (Shropshire Site and Monument Records No: n/n, accessed 07/08/2009, Shropshire Magazine April 1965, accessed 05/08/2009 and Adams, 2007). Sylvan Close, part of a housing estate, now occupies the former Colliers Side Quarry site. A study carried out at the site, by Ove Arup Company in 1987, suggests that the former limestone workings at Collier Side Quarry were approximately 40 to 45 metres deep in places (Ironbridge Gorge Museum Trust, Ove Arup *et al,* 1987).

Maps from the 19th and 20th centuries of the former Colliers Side Quarry area indicate the immense size of the mining activities taking place underground. These maps also show associated buildings, tramways and canal. In 1997, Sir William Halcrow and Partners Ltd carried out a survey for Shropshire County Council on the former Colliers Side Quarry site; their report indicated a similar depth of limestone workings, as shown in the first report of 1987 (Adams, 1997 and Ironbridge Gorge Museum Trust, accessed 05/08/2009).

6.9.4 Jackie Parr Hole Quarry NGR: SJ732161

Jackie Parr Hole Quarry is said to have been named after a local man called John Parr, who is believed to have lived in an abandoned kiln. The quarry was situated to the west of Limekiln Lane, Lilleshall. The southern part of the quarry was backfilled in the 1960s, while the northern end of the quarry was used as landfill site; nothing remains visible today. The stone quarried from Jackie Parr Hole Quarry was lower red or hydraulic limestone, produced from a unique natural raw material which is impervious to water. This type of limestone was used in the construction of the Liverpool Docks; because of its chemical properties it was used as a hydraulic binder in mortar before the development of the original Portland cement. By 1858, Jackie Parr Hole Quarry was exhausted (Robert Galloway, Lilleshall Site Visit, 22/08/2009, Shropshire Archives, Lilleshall Records, British Geological Survey: The Economic Importance of Minerals to the UK, www.mineralsUK.com, accessed 06/04/2011 and Adams, 2007).

6.9.5 Willmoor Mine (Sour Leasow Pit) NGR: SJ736715

In 1846, The Lilleshall Company discovered new limestone deposits south of Willmoor Bridge; this site would be known as Willmoor Mine. The mine was previously referred to as Sump Leasow or Sour Leasow Mine, near Willmoor Lane. Willmoor Mine opened, or possibly re-opened, in 1858, to work the hydraulic grey limestone seam; one of these seams had measured 4.40 metres thick (Ironbridge Gorge Museum Trust, Ove Arup *et al,* 1987 and Armstrong, 1984). Previous to the mine re-opening in 1858, a survey of the mine showed that two shafts had been sunk to a depth of approximately 60 meters; these subsequently collapsed. Willmoor Mine ceased operation in 1882; it had been an active mine for 24 years producing over 185,000 tons of limestone in its life-time. It seems that the Lilleshall Company could obtain cheaper limestone from Wenlock Edge Quarries and therefore it became uneconomical to mine limestone from Willmoor Mine (Ironbridge Gorge Museum Trust, Ove Arup, *et al,* 1987, Shropshire Record Office, Sutherland Collection, document no, D593/M/17/5, accessed 10/11/2009 and Adam, 2007).

6.9.6 Presthope Quarry, Wenlock Edge NGR: SJ575986

In the early 1700's, near to Presthope, there were indications that quarrying and lime burning had been carried out. Likewise, in 1846 quarrying and lime burning had taken place on the north-east of Hilltop, both sides of the B4371, Church Stretton to Much Wenlock road. This has now been quarried away by the Lilleshall Company. By the 1860s, the Lilleshall Company's limestone resources around Lilleshall Village and Church Aston were becoming exhausted. The company mines that still operated were subject to flooding that could not be controlled (Gale *et al,* 1979 and Holmes, 1978). In April 1862, Earl Granville Leveson-Gower of Granville on behalf of the company, leased 28 acres of land between Hilltop and Presthope at Wenlock Edge, from Moses George Benson of Lutwyche Hall. In the 1860s, production at Presthope Quarries considerably increased so that in 1864 the Much Wenlock railway needed to construct a branch line to transport the processed lime. In 1876, the lease of the 28 acres of land was renewed and agreed under the original terms. The Lilleshall Company re-entered negotiation in 1896 with the Benson Family for the renewal of the lease. The lessor stated that the rent required was £500 per year with extra royalties of 4d per ton of limestone quarried and 12d per cartload or 56 bushels. Further negotiation between the two parties saw the rent set at £200 per year, but the royalties stayed the same. A contract was signed on 8th June, 1896 renewing the quarry lease for a further 14 years. By 1900 production at Presthope Quarries had reached 1,000 tons per week. In December 1910 the lease of the Presthope Quarry was requested for a further 7 years. This was endorsed in March 1911 by the Benson Family; however, the rent was reduced to £140 per year. Even so, at a Lilleshall company board meeting on January 29th 1915, it was decided that the lease would be surrendered on 25th March 1916 and written notice was given to the landowners, the Benson Family. The major factor in the Lilleshall Company surrendering the lease was the lack of manpower to work the quarries, caused by the First World War. Ultimately, in 1917, the steel railway lines that had connected Presthope with Much Wenlock were lifted and used to aid the war effort (Shropshire Archives, Lilleshall Records, 809/6/1, 809/6/2 809/6/4 and 809/6/5, accessed 05/08/2009 and Yates 2008).

6.10 Wenlock Edge Limestone Workings

6.10.1 Lea Quarries NGR: SJ594982

There have been a number of quarries on Wenlock Edge over the centuries, during which time, some have been amalgamated. The Lea Quarries, were situated to the north-west and south-west of the B4371on the Much Wenlock to Church Stretton Road, both of which were operated by Ridge Limestone Ltd. The proprietor of Lea Farm in 1841 was a Mr Richard Milner; a quarry on his land shown in the tithe map of 1848, could have been operated by this farmer (Williams, 1990). It was not until, 1943 some two hundred years later, that major quarrying started to take place on this site. The site of the quarry was then owned by Messrs Cross and Sons Ltd, who quarried approximately 5 acres of land on which they installed crushing and milling plants, in order to produce crushed limestone for horticultural fertiliser and employed some 65 quarry workers. Originally, from the total production of limestone quarried at Wenlock Edge, 80% of the limestone was destined for agricultural purposes, but by 1947 this had dropped to only 20%. In 1947, Messrs Cross and Sons Ltd were producing 200 tons of limestone per week in the winter period; the larger portion of this was used in construction. In the spring and summer months of the same year, productivity was up by 50% to 300 tons per week; a boost in productivity meant an additional nine men were employed (The Quarry Managers' Journal, 1968). In 1948, Much Wenlock Council received an application from the proprietors of the Lea Quarries, for renewal of their quarrying licence. This was granted for the quarrying of 210 acres of land with the provision that the quarrying should not encroach within 27.43 metres of Wenlock Edge (Shropshire Archives, Ref DA6/101: Much Wenlock Council Minutes, accessed 23/04/2010, document no 958, accessed 23/04/2010 and Williams, 1990).

In 1957, The Ridge Limestone Ltd purchased the quarries and extended their workings into Westwood Quarry, thus amalgamating the two sites into one now known as Lea Quarries. Between 1959 and 1961, Ridge Limestone Ltd started the process of modernising Lea Quarries; they installed a primary processing, secondary crushing, screening plant, tertiary crushing, grinding plant, loading and feeding bins. The new technology installed increased production to approximately 1,000 tons of burnt lime and 3,000 tons of limestone per week (Williams, 1990). Illustrations below show the plant machinery at Lea Quarry:

Figure 6.17: Primary Crusher (Source: Birmingham Archaeology, 2009.

Figure 6.18: Primary Crusher. (Source: Birmingham Archaeology, 2009).

Figure 6.19: Primary Crusher showing chain screen. (Source: Birmingham Archaeology, 2009).

Figure 6.20: Primary Crusher with intake hopper with chain screen (Source: The Quarry Managers Journal, 1968).

Figure 6.21: High Intensity Screen handling scalping plant (Source: The Quarry Managers Journal, 1968).

Figure 6.22: West Screen Conveyor. (Source: Birmingham Archaeology, 2009).

Figure 6.23: View of Hoppers. (Source: Birmingham Archaeology, 2009).

Figure 6.24: West Screen Conveyor (Source: Birmingham Archaeology, 2009).

Figure 6.25: View of Hoppers. (Source: Birmingham Archaeology, 2009).

From 1961 to 1962, Ridge Limestone Ltd applied, and obtained, new quarrying rights, with the proviso that the works did not impinge within 15.24 metres of the B4371. This excluded a small parcel of land north of the B4371, owned by another party; quarrying activities commenced to the area near to this excluded site. In 1968, the main market of Lea Quarries was in the manufacturing of concrete blocks and brick making. There were a number of industrial markets for limestone within 30 to 35 miles radius of Lea Quarries, which made the production of these blocks and bricks economically viable. In the same year major new housing developments took place in Droitwich and Telford, formally Dawley, thus boosting productivity at Lea Quarries for limestone, strengthening the finances of the company (Shropshire Archive, document no 61/2250, accessed 23/04/2010, Williams, 1997 and Salop County Council, 1997).

By 1968, Lea Quarry was producing approximately 5,000 to 5,800 tons of limestone per week or approximately 279,000 tons per year, which was principally used for roads; even so, this figure was amplified by limestone quarried from the adjoining Coates Quarry which was also owned by Ridge Limestone Ltd. Lea Quarry grinding mills could produce approximately 30,000 tons per year. In 1977, some nine years later, production had reached approximately 600,000 tons per year. The breakdown of the production figures for that year shows: 80% aggregates, 10% crushed limestone for concrete merchandise, 8% agricultural lime and 2% for fluxing stone (Williams, 1997, Trinder, 2000 and Salop County Council, 1997). Again in 1988 production had increased to 700,000 tons per year. Summarising Lea Quarries production levels show an increase of a 100% by 1988, some nine years later. However in 1988, productivity only increased by 10%, which indicates a slowing down in demand for limestone. In 1988, Lea Quarry employed 20 quarrymen using lorries to transport the quarried limestone to the processing plants, following which the processed limestone was taken to all parts of the country. The B4371 which Lea Quarry had to use, from Much Wenlock to Church Stretton is a narrow road and not suitable or constructed for such heavy industrial traffic (Williams, 1997 and Salop County Council, 1979).

The scale of productivity at Lea Quarry north-west workings also relied on stone quarried from Westwood Quarry and the majority of Hayes Quarry. In the mid-1970s, at Lea Quarry there were landslips caused by water lubricating between the clay and limestone strata, causing large blocks of limestone to slide down after the quarrying had finished (Williams, 1997 and Salop County Council, 1979). In 1992 Lea Quarry suffered a slump in demand for its limestone, thus causing some employees to be made redundant. The earliest workings at Lea Quarry were now exhausted after centuries of extraction, leaving only the plant machinery buildings extant. However, quarrying continued to take place at Lea Quarry south, opposite to this original site. Sometime in the mid-1990s Ridge Limestone Ltd sold Lea Quarry to Bardon Aggregates Ltd who operated the site until it was mothballed in 2007. It has been suggested that Lea Quarry south has enough limestone reserves to last until 2042 (Robert Galloway, Wenlock Edge Site Visit, 14/04/2010, Williams, 1997 and Salop County Council, 1979).

Even in the late 19th century discussions were taking place concerning quarrying and the environment at Wenlock Edge (Williams, 1997). In 2000, Shropshire County Council informed Bardon Aggregates Ltd, owners of Lea Quarry that no further permission would be given for the expansion of quarrying. The reason given was that the quarrying industry was now considered as not being environmentally friendly; also no application for re-location of the quarry would be considered. Even if any future expansion or re-location of the Lea Quarry had been considered, moving all the plant equipment would have been too costly, so making it a financially unsustainable operation (Salop County Council, 1979 and Birmingham Archaeology, 2009).

After the mothballing of Lea Quarry in 2007, the future of the former quarry site has been debated; as with any former industrial site, if left, nature will re-claim it. The majority of Wenlock Edge is designated a Site of Special Scientific Interest, so that significant fauna and flora may be recorded as having re-colonized the site.

Another point to consider would be that the former Lea Quarry limestone site with all its plant equipment should become a monument to Wenlock Edge's industrial past (Robert Galloway, Wenlock Edge Site Visit, 14/04/2010, Salop County Council, 1979 and Birmingham Archaeology, 2009).

6.10.2 Coates Quarry NGR: SJ605993

Coates Quarry is the nearest quarry located to Much Wenlock Town. In 1770, just between Coates Farm and Blakeway Lane was Lime Pit Field, located near the present quarry entrance, and so possibly the original quarry site before the enclosure of Westward Common in 1806-1807. Access to Coates Quarry was by way of the B4371, the Much Wenlock to Church Stretton Road. Although quarrying was taking place in the early1800s, as there is an absence of data including census information it has not been possible to attach names to specific quarry locations prior to 1841. In 1841, Lord Forester rented out Lime Pit Field, together with limekilns, but no names were provided (Williams, 1997, Salop County Council, 1979 and Birmingham Archaeology, 2009). The tithe map of 1847 indicates a small disused quarry near the summit of Wenlock Edge, north-west of Coates Quarry. In 1848 the Lime Pit Field was being rented and quarried by Mr William Jukes for £5: 5 shilling, per half year. Thomas Jukes, recorded as farming Coates Farm in 1851-1852, was also known to be quarrying limestone and producing lime. It is possible that Thomas and William Jukes were brothers, but no record could substantiate this. It is recorded that in 1857 the Gas Works in Barrow Street, Much Wenlock, was purchasing lime quarried by Mr William Jukes for 10 shillings and 6 pence per cart load. In 1859 William Jukes is recorded, in Slaters Directory of Shropshire, as being a lime burner at Coates Quarry (Historical Directories. www.historicaldirectories.org, accessed 03/06/2010).

In 1863, Mr Thomas Jukes took control of Coates Quarry and is recorded in the historical directories of Shropshire as being the quarry owner, until sometime in 1888 when Mr William Reynolds is shown as having gained ownership of this quarry. From 1890-1907, Mr George Meakin, was quarry master at Coates Quarry, whether he was the owner, or not, is unclear (Historical Directories. www.historicaldirectories.org, accessed 03/06/2010, Williams, 1997 and Shropshire Archives, Porter's Directories, accessed 23/04/2010). By 1910 Coates Quarry was in the ownership of Mr George Lloyd; at this time there were five limekilns on the site. In 1914, Coates Quarry was being worked by Mr Edwin Sankely, who paid £10 per year to George Lloyd until his death in 1924. After his death all rent and royalties of either 6d or 4d per tonnage, dependent on the quality of limestone produced, passed to his estate. In 1924, George Lloyd's executors put Coates Quarry up for sale, yet the date when the quarry was actually sold is not clear. Kelly's Directories state that in 1926, 1929 and 1934 Edwin Sankely was the owner of Coates Quarry (Salop County Council, 1979, Williams, 1997, Victoria County History of Shropshire, www.british-history.ac.uk, accessed 26/03/2010 and Historical Directories, www.historicaldirectories.org, accessed 03/06/2010).

Coates Quarry was sold sometime in 1934, yet who purchased it is unclear, however in 1933 a Captain Kane bought Westwood Quarry and it is a possibility that he also acquired Coates Quarry at the same time. Coates Quarry had increased its limekilns from five to six and was still producing limestone at this site in the early years of the Second World War. In 1944, Kane sold Coates Quarry to Kalkyne Mining Company who immediately set about modernizing the Coates Quarry by installing crushing and milling plants (William, 1997 and Salop County Council, 1979). Kalkyne Mining Company modernized Coates Quarry, only to sell it in 1945 to Messrs Peddigrew, Jones and Horner. The new proprietors of Coates Quarry employed 20 men at the height of production, working day and night shifts producing approximately 1,200 tons of limestone per week. In the late 1940s records show that Coates Quarry was under the ownership of the Ridge Limestone Ltd; also the owners of Lea Quarry. In the early 1950s, Ridge Limestone Ltd applied for a quarry extension and this was granted in 1953, with conditions attached specifying that quarrying would not be allowed in certain areas north and south-west of the site. An area of 44 acres was covered under consent No. 897 (Salop County Council, 1979). The 1950s saw Coates Quarry at the height of its production, working day and night, 7 days a week, gradually increasing the amount of limestone quarried and by 1955, approximately 70,000 tons of limestone per year was being produced. With the increase in productivity more modern crushing and milling plants were installed to cope with the production. With the continuous operation of quarrying at Coates Quarry, local residents started to complain about the constant noise and dust; particularly the blasting operations at weekends. Residents also complained about the increased quarry traffic on the B4371 and adjoining roads and subsequent additional noise associated with the heavy transport (Salop County Council, 1979 and William, 1997).

In 1963, production at Coates Quarry had reached approximately 100,000 tons of limestone per year; however in the same year Ridge

Limestone Ltd began to reduce quarrying at the site. By 1970, quarrying at Coates Quarry ceased and all the quarrying machinery was dismantled, demolished or sold off (Robert Galloway, Wenlock Edge Site Visit, 14/04/2010, Salop County Council, 1979, and Williams, 1997). The disused Coates Quarry was now being utilised by Nature Fuels Ltd, a company researching into solar energy, who obtained temporary planning permission until 1980. Then Mucklow and Sons Ltd were granted planning permission to use the former quarry site until 1986, with the condition that they did not re-generate the quarry. They used the south part of the quarry to manufacture and store concrete.

In 2010, the former Coates Quarry was used as a store for larch lap fencing, which they distributed. The only machinery left on the site of the former Coates Quarry is the primary crusher, now in a state of decay.

6.10.3 Knowles Quarry, NGR: SJ585997

Knowles Quarry, as it is known today, was once called Presthope and Limerock; in the earliest reference to this site Mr John Mason is recorded as quarrying in 1714. The next recorded working of the quarry, in 1838, was by a Mr Edward Rogers, until a Mr Edwin Hayward acquired the quarry in 1846; he leased the quarry from the Earl of Bradford. Around the same time a Francis Price who occupied The Knowl (Knowle Cottage), near to Knowles Quarry, could possibly have been working at the quarry. In 1868, Mr Edwin Hayward is recorded in Slater's Royal National Commercial Directory as being a lime burner at Presthope (Knowles Quarry) (Historical Directories, www.historicaldirectories.org, accessed 03/06/2010, Shropshire Archives, 1846 tithe map of Much Wenlock, and document no. Ref: DA6, Wenlock Rate Book, 1838, accessed 23/04/2010).

In 1880, Mr Edwin Hayward was succeeded by a Mrs Letitia Hayward as the lessee of Knowles Quarry. On the 1881 census Mrs Letitia Hayward, 74 years old, is recorded as a widow (presumably) of Edwin Hayward. On the same census Letitia Hayward's residence is given as The Knowl (Knowle Cottage) and her recorded occupation is quarry manageress of Knowles Quarry. Letitia Hayward employed her sons, Henry 45, Samuel 43, George 36 years of age and her grandson 27 years old Edward Hayward, as quarrymen; she is also shown to have had a granddaughter called Mary Cottrel (1881 England Census, www.search.ancestry.co.uk, accessed 14/09/2010 and Williams, 1997).

How long Letitia Hayward was in control of Knowles Quarry is not known, because in the 1890s her sons Henry and George Hayward finished their family association with Knowles Quarry. In 1891, Mr William Hayes was recorded as being a lime merchant in Much Wenlock and in 1895 he took over the lease of Knowles Quarry. Five years later in 1900, William Hayes and Son were the proprietors of Knowles Quarry and are recorded as using tramways to transport the quarried stone from the quarry face. In 1911, Mr Alfred Lucas Hayes renewed the lease on Knowles Quarry; he traded under the name of William Hayes and Son, quarrying the site for another 14 years. This limestone was used to construct local buildings and all over Shropshire. Alfred Lucas Hayes paid the Earl of Bradford £12 per year and 3d per ton as a royalty on all limestone excavated from Knowles Quarry. Included in the lease were the stables, the power house and cottage (Shropshire Archives, Lilleshall Records, 809/6/1, 809/6/2, 809/6/4 and 809/6/5, accessed 05/08/2009, Historical Directories, www.historcaldirectories.org, accessed 03/06/2010, Shropshire Archives, Kelly's Directory of Shropshire, accessed 11/06/2010 and Williams, 1997). Compared with the £140 per year rent paid in 1910 to Mr M. G. Benson, for the Presthope Quarry at Wenlock Edge by the Lilleshall Company, the rent of £12 per year paid in 1911 to the Earl of Bradford, for Knowles Quarry Wenlock Edge paid by Alfred Lucas Hayes was a comparatively small amount. The amount of reserves left and therefore life-expectancy, or difficulty with road access, perhaps played a major part in deciding the value of these rents.

In 1919, The Earl of Bradford sold Knowles Quarry, the cottage and the lease agreement, previously held by Alfred Lucas Hayes, to a Mr George Newton for £400. In the early 1920s it is shown that Alfred Lucas Hayes surrendered his lease for Knowles Quarry, although, the 1926 Kelly's Directory records him as still having a business in Much Wenlock. It is noted that from 1919 to 1921 a Mr J. Hensaw was carrying out blasting operations at Knowles Quarry; could he have taken on the lease of the quarry after Alfred Lucas Hayes? In 1921, the Much Wenlock Town Council instructed the Town Clerk to send Mr J. Hensaw a letter stating that his blasting operations at Knowles Quarry were getting dangerously negligent. The Town Council were concerned about the proximity of the Knowles Quarry causing disruption to transport on the B4371, Church Stretton to Much Wenlock road. It was requested that for any future blasting operations, men with red flags should be on the road giving warning to traffic and the public (Shropshire Archives, Kelly's Directories, accessed 11/06/2010, documents no. Ref: DA/130, Much Wenlock Town Clerk, accessed 23/04/2010 and Adams, 2007).

When Knowles Quarry was sold to Mr C. R. H. Kane is not clear, possibly in the mid-1920s. However in Kelly's Directories 1934 and 1937,

Mr Kane is recorded as being the proprietor of Knowles Quarry. In addition, to Knowles Limestone Quarry, Mr Kane had Plough and Stretton Quarries from which he also extracted limestone (Shropshire Archives, Kelly's Directories, accessed 11/06/2010 and Holmes, 1987).

In the early 1950s, Mr Kane sold Knowles Quarry to The Ridge Limestone Ltd, who continued to quarry under policy no. 4254, which allowed an area joining Hayes Quarry to be worked out. In 1954, the County Council granted permission under policy 9114 for a combined area of 47 acres to be quarried north and south of the B4371, Church Stretton to Much Wenlock road. Ridge Limestone Ltd was now the major quarry owners in the Wenlock Edge area, combining Knowles and Lea Quarries; the new site would be known as Lea Quarry. In 1991, The National Trust took control of the former Knowles Quarry, once a part of Lea Quarries, and installed their warden in Knowles Cottage. Today, the former Knowles Quarry is a Site of Special Scientific Interest (Salop County Council, 1979 and Williams, 1997).

6.10.4 Hayes Quarry, NGR: SJ604994

On a map of 1889 a small quarry appears near the main road; this is the first recorded evidence of the quarry (Edina Digimap Historic, accessed 23/09/2010). This quarry site was first known as Oakley's Quarry, until the mid-1920s, then as Cox's Rock Quarry and latterly as Wenlock or Hayes Quarry. There is very little evidence of quarrying taking place at this site until after 1918 when serious limestone extraction began. In the early 1920s a Mr George Lloyd owned Coates and Hayes Quarries, which consisted of 7.5 acres. In 1923 George Lloyd leased Hayes Quarry to Mr John, Edwin, Frederick and William Oakley for a period of 5 years to quarry limestone. The Oakley brothers paid George Lloyd a rent of £22:5:00 shilling per year, plus royalties of 6d per ton of burned lime and 4d a ton of limestone quarried (Williams, 1979).

The Oakley brothers' surrendered the lease of Hayes Quarry in 1925; the reason given by John Oakley was the decline in the limestone industry. Hayes Quarry was put up for auction by the executors of the late George Lloyd, but did not sell. In March of 1926, Hales Quarry was purchased by Gladys Cox for approximately £250. However some months later in 1926, Hayes Quarry was being operated by Alfred Lucas Hayes, trading under William Hayes and Sons Ltd who had in the early 1920s surrendered the lease of Knowles Quarry. So apparently Gladys Cox must have sold the Hayes Quarry soon after purchase. Hayes Quarry had 5 kilns and in 1929 an additional one was built show that business was good; in 1935 evidence indicates that the lime burning facilities were increased again to a total of nine kilns. Additionally, in 1935 a loading platform was built together with a grinding mill and storage shed. In the late 1950s at Hayes Quarry tramways were still being used to transport quarried limestone from the quarry to the lime kilns; the horses still pulled the tubs from the quarry face to the loading area. Hayes Quarry employed twelve quarrymen who worked the quarry face, also breaking up large rocks with sledge hammers (The Quarry Managers' Journal, 1968, Shropshire Archives, documents no, Ref: DA6/100/8, Wenlock Council Minute Book, 1953, accessed 23//04/2010 and Williams, 1997).

A major customer of William Hayes and Son was the Round Oak Steel Works, Brierley Hill, but this started to decline in the 1950s. The decline was not helped by the high cost of fuel, needed to produce burnt lime. In 1953, Wenlock Council received an application to extend the quarrying operations at Hayes Quarry, which was granted under policy no. 4254. This would allow productivity to increase between 12,000-14,000 tons per year. At Hayes Quarry in the mid-1950s a Jaw Crusher and Hammer Mill was installed to process lime; the plant could produce approximately 5 tons of ground lime per hour (Shropshire Archive, documents no. 4254, Ref DA/6/100/8 Wenlock Council Minutes, 1953, accessed 23/04/2010 and Williams, 1997).

Hayes Quarry in the late 1950s to early 1960s was at its pinnacle, producing approximately 15,000 tons of limestone per year; spring and autumn being the quarry most productive periods. The quarry virtually lay idle throughout the winter months. It seems that sometime in 1964, Hayes Quarry changed hands; Ridge Limestone Ltd being the main quarry operators of Wenlock Edge at the time it is more than likely that the quarry was amalgamated with Lea Quarry (Salop County Council, 1979 and Williams, 1997).

6.11 Discussion

The research into both Lilleshall Company and Wenlock Edge Quarries historical and archaeological background has been intense. There are similarities which have come from researching the background of Lilleshall and Wenlock Edge industrial past. For example with the dissolution of the monasteries by King Henry VIII in 1536-1541, some people with money were able to acquire the monastic lands. Lilleshall Abbey lands were purchased by the Leveson family in 1539; Much Wenlock Priory lands were purchased by a group of people; Lawrence Benthall, Sir Basil Brookes, James Clifford and John Weld in 1540 to add to their already existing estates. These groups of people knew of the potential mineral deposits beneath the monastic lands and the money that could be made from the extraction of the limestone. Nevertheless there is one major difference between Lilleshall and the Wenlock Edge; Wenlock Edge has always been a place of outstanding beauty. However the natural beauty of Wenlock Edge did not stop quarrying taking place from the earliest time to build Much Wenlock Priory and town. One could suggest that in the 21st century, that it is important to preserve the natural environment from the effects of quarrying. However the industrial past has played its part in the economic growth of Lilleshall, Wenlock Edge and Shropshire. So the industrial man-made landscape has become linked with the natural beauty and geology of Lilleshall, but more so Wenlock Edge.

The industrial enterprise of 'limestone quarrying' was started by Sir John Leveson, later Earl Gower and the Gilbert brothers in the early 1750s; then in 1802 became the Lilleshall Partnership (Company). 1802 is when the limestone quarrying really commenced at Lilleshall and by 1846 there are indications that white limestone suitable for fluxing had been exhausted across the estate. However the Lilleshall Company still extracted hydraulic grey limestone until the 1880s. In the 1980s the Lilleshall Company finally ceased its connections with Shropshire.

The progressive ways of transporting quarried limestone was: by people, horse-waggons, tramways, turnpikes, canals, railways and road; however some of the quarry locations made the moving of quarried limestone difficult. As well as transport, specific important tradesmen were also associated with the quarrying industry, such as blacksmiths and doctors. Blacksmiths maintained and made parts for the different machinery; the blacksmiths also sharped the tools used by the quarry workers either at the quarry face or for dressing the quarried stone. All the quarry workers may have needed looking after by a doctor, who was paid one penny per week, which came out of the workers' wages. During the extent of the research for this book, Mrs Letitia Hayward in 1881 at Knowles Quarry, Wenlock Edge at the age of 74 years, is the only woman noted as a quarry manager. It is remarkable that a woman of 74 year old was doing such a job in the 19th century. The B4371, Wenlock Edge to Church Stretton Road is a rural road not made for heavy lorries from the once active quarrying industry of the area. As with all the quarries I have researched the movement of the processed limestone today with HGVs causing problems, such as dust and noises are just two examples for the local areas and people.

From site visits to Lea and Coates Quarries I was able to see the difference in the standard of machinery. For example if one looks at the pictures taken of the primary crushers at both sites, firstly Lea Quarry the equipment is quite modern and secondly Coates Quarry is not. However the design of the machinery is the same. I acknowledge that Coates Quarry closed in the 1970s and Lea Quarry was mothballed 2007; however it proves an excellent example of machinery used over a thirty-seven year period. Also the figure 16 picture of Coates Quarry gives idea of the amount of limestone extracted from just one site. There were another three large sites on Wenlock Edge and the five sites at Lilleshall. The other side of the limestone quarrying industry was the workforce what was the quarryman's working life; what was their life expectancy and their families. It has been established that in the 19th century most quarry workers lived in cottages located near to the quarries, owned by the quarry owners. So everything was kept reasonably local therefore benefiting the local economy. As far back as the 1880s concerns were being raised about the visual intrusion, damage to the landscape by smoke and noise from blasting operation, dust and traffic around Wenlock Edge; however this was a national problem as well as a local one. These early concerns were the start of environmental issues that would face quarrying up till the present day.

The principal difference between Lilleshall and Wenlock Edge is that of location; Lilleshall is neighbouring farmland and houses, so that one of the main problems that Lilleshall has is the former limestone mines which occasional subside. Wenlock Edge on the other hand never had limestone mines; however in the 1970s there were two landslips. Today, Wenlock Edge provides stunning views of Shropshire; Wenlock Edge also has a scarred landscape from it limestone quarrying past. Now the beauty and the industrial past of Wenlock Edge are combined the limestone escarpment now provides excellent conditions for many rare flowers; also it supports ancient woodlands and is famous for its geology. The woodlands now provide some excellent walks, riding and cycling routes. From the edge of the woodland one can see the former quarries. However the edges of the quarries drop straight down and are inaccurately fenced; the danger is still there for people and livestock.

6.12 Conclusion

The research into Lilleshall Quarries, Mines and Wenlock Edge Quarries showed some similarities and differences. With the dissolution of the monasteries in 15361541 people with wealth, influence and money could purchase the monastic lands and exploit the limestone from the land as in the case of Lilleshall and Wenlock Edge. At Lilleshall quarrying and mining for limestone took place, in the case of Wenlock Edge only quarry for limestone happened. Mining at Lilleshall was a dangerous occupation and men were killed when the roof of the limestone mine collapsed on them. There were no mines at Wenlock Edge; however because of the quarry activities two landslips did happen, but there were no fatal industrial accidents recorded.

The Gilbert brother's partnership with the Lilleshall Company provided the knowledge for canal planning, buildings and legal advice. They were fundamental in the construction of the first private canal at Donnington Wood, Lilleshall and in the building of their own private railway. Also with the coming of the National Railway System in the 1840's both the canals and railways brought considerably changes to the industrial scene. It also helped in the transporting of limestone from Lilleshall and Wenlock Edge. Canal transport was very good, yet the railways improved on this and were able to transport goods even further to place the canals could not reach, thereby increasing financial and production levels for The Lilleshall Company. Towards the end of the 19th century, Earl Gower and Company built a hospital for his workers and a school for their children at St. Georges.

Records at Shropshire Archive relating to Lilleshall, Wenlock Edge and the Ironbridge Gorge Museum Trust Collection of Documents Relating to The Lilleshall Iron & Steel Company Ltd have assisted in the research of these two sites.

CHAPTER 7

CONCLUSION

At the outset of this concluding chapter, I shall re-state the key aims of this research and discuss them in turn.

1. To explore the significance of the Shropshire quarrying industry: Shropshire's geological diversity contributed to the development of the bulk mineral industry. It is accepted that ten geological periods can be found in Shropshire, whereas in comparison Snowdonia has only three; the combined remainder of the country has just twelve such periods (Toghill, 2006, p.27). Shropshire is an ideal place to study the quarrying industry. When the Romans came to Britain in 43 AD, it did not take them long to see the significance and potential of the minerals to be found, for instance, in Llanymynech Hill and at Wenlock Edge; these were subsequently used at Wroxeter as lime mortar and for mosaics. Sandstone was also quarried in the county at this period from a number of sources as construction material for buildings (Roger White, per comm, 04/08/2011). Shropshire minerals, particularly limestone, were firstly produced as agricultural fertiliser and widely used throughout the county. The monks at Lilleshall Abbey and Wenlock Priory quarried limestone block to construct their monastic buildings. The significance of the bulk mineral quarrying can be seen across Shropshire; for example, one only has to look at the buildings and bridges in Shrewsbury. Entrepreneurs in the 1500's saw and grasped the opportunity to purchase the former monastic lands, because they knew of the potential of the minerals in the ground. Later businessmen would, over the period under research, develop Shropshire's quarrying industry until its decline in the early 20th century.

2. To understand the social impact of the quarrying industry on the workforce of the region: Initially mineral extractions in Shropshire were undertaken by the monks for monastic buildings although small family operated quarries also existed. Coates Farm, Wenlock Edge, is the original site of a small quarry worked by the farm owner for agricultural fertiliser for his land. Additionally at Llanymynech, evidence gained from late 18th

century wills, drawn up by local people, show owners of small quarries passing quarries and tools on to family members. This gives an indication of small family owned and operated quarries before the commencement of large commercial quarrying operations. Small quarry owners were the most affected by the Enclosure Act of 1818, the provisions of which resulted in the loss of all their lands, quarries and livelihoods. Initially, records show that as early as 1714 on Farley Common, Much Wenlock, small quarrying operations for limestone had been taking place. Yet, following this a hundred years later, as an effect of the Enclosure Act in 1818, 102 acres at Farley Common were enclosed, so bringing to an end the work of all of the small quarry owners. Thus, the entrepreneurs and large landowners would be the only people to benefit from the Enclosure Act 1818. The purchase of land did not stop after the Enclosure Act, 1818. In Much Wenlock an example of large landowner was Lord Forester, who owned 194 acres in 1848. Ten years later in 1858 this had been increased by the purchase of another 827 acres, bringing the total of land owned to 1,021 acres in the Much Wenlock area. So, from this one Act of Parliament, one can see how small, local quarry owners were affected. They lost their small quarries, which had been absorbed by the entrepreneurs and large landowners for their own benefit. From these small initial quarry sites, larger quarries would be developed. Small quarry owners were left with common heathland which had no commercial value. So, from this one Act of Parliament the small quarry owner operators had lost everything, their livelihoods and their independence (Shropshire Archives, 1224, box 108 deeds and documents 1681/42/16, accessed 23/04/2014).

Working life for people in Shropshire, a mainly rural county, was in agriculture; although there were small family owned quarries, as well as Abraham Darby's blast furnaces at Ironbridge, which was founded in 1709 prior to the beginning of the major industrial quarrying operations. By the end of the 17th century, limestone quarrying in Lilleshall was well established. At this time in these quarries self-employed lime carriers who worked long days. However, the 19th century would be the height of limestone production in Shropshire's quarrying industry, which in turn saw the rise in fortunes of the entrepreneurs. At the same time the industrial economy of Shropshire was thriving. Throughout these times economic benefit from the growing quarrying industry was slow to reach its working people. The quarry worker's days were long, particularly at Brown Clee Hill where the quarry workers started from home at 4am to begin work at 6am. An alternative to this would be to find lodging, or sleep rough, which could be dangerous if the weather, was cold. This thus caused the labourer to be away from their family for six out of seven days in the week.

In 1810, The Lilleshall Company built a block of houses for their workers known as “Barrack Houses”. At some point between the late 19th and early 20th centuries The Brown Clee Hill Stone Quarry Company built a row of houses for their expanding workforce. Accommodation built by the quarry proprietors for their quarry workers had its draw backs, because the quarry workers were always under threat of eviction from their homes if they became dismissed from employment, thus putting their families at risk of homelessness and destitution. To keep a roof over their own, and their family’s heads, the quarry workers would have to work double shifts excavating stone in the summer, to make up in advance for time which may be lost, because of bad weather in the winter months. This was required by the quarry owners to enable the processing of quarried stone to continue to be worked on, during the winter months, by the other workforce.

At the Llanymynech limestone quarries, a piece-rate system was operated with the gangs of quarry workers working a six-day week; if they did not work, they were not paid. Similarly, to the working practices at Brown Clee Hill, it was also common practice at Llanymynech for the gangs of quarry workers to have to work double shifts in the good weather to make up for the bad weather in the winter months. This shows what the life of the quarry face workers was like; hard work, long hours in the summer months and being laid off in the winter months when the weather was bad. With the threat of eviction always hanging over the quarry workers’ and their families’ heads, they had no choice but to work the long hours. The job of a quarry worker was dangerous and by working long hours this danger would be increased by tiredness. In the 19th and early 20th centuries there was no legislation to stop the evictions of the quarry workers and their families; it was quarry owners who had the upper hand. This gives one an insight into the working lives of the quarry worker who extracted the minerals, worked long hours for little money, and needed to keep a house for their family. It should be remembered that along with the people lucky enough to have full-time quarrying jobs, there were people employed on a casual basis. They would be living day to day with little money to feed themselves and their families.

3. To explore the historical background of the individual quarry sites, so as to provide an understanding of their industrial growth: It was important to establish the foundations of the major quarrying industries in Shropshire. Visiting archives in Shropshire, Staffordshire and The National Library of

Wales provided the historical documentation required. The Ironbridge Gorge Trust was another place where historical records of the Lilleshall Collection are held and provided the background to The Lilleshall Company. These historical documents give one an insight into the initial activities and importance of quarrying in Shropshire, from individuals to the later entrepreneurs, who developed the quarrying industry in the county. Through my research I was able to establish the people who might best offer a wider background to the operating history of the quarrying companies, and from whom I should seek interviews. Additionally, the Chief Planning Officer of Shropshire County Council, and local councillors proved to be a further resource to interview. These are an example of the steps undertaken by myself in order to establish a complete historical and archaeological record of the Shropshire quarrying industry.

The earliest researched landowner to start to develop the former Lilleshall Abbey lands was James Leveson who purchased the monastic lands in December 1539. His family would improve the former monastic lands by draining and creating causeways across it, in order to make easier access for the development of the quarries and the eventual transport of the quarried limestone. In the 16th and 17th centuries, there then followed the social upheavals of the Reformation and the English Civil War, after which in the 18th and 19th centuries there was a period of economic growth and development. This enabled the entrepreneurs to explore and expand the uptake of their limestone into other industrial fields, such as contact with the iron masters who would need more limestone for the flux in their blast furnaces. So consequently, the limestone quarrying industry needed to expand and find new sources of limestone in Lilleshall quarries and mines. My research showed the dangers that the quarry worker faced, however the danger would be increased when working in mines and mining for limestone. This expansion also happened at the Wenlock Edge Quarries. Growth of the mineral industry continued into the early 20th century. Through my research into other quarries in the county, I was able to establish that, with the coming of the Industrial Revolution, minerals such as dolerite and greywacke were utilised in road making. This was a very productive period for the mineral industry of Shropshire.

4. To investigate 19th century legislation and its applicability to the quarrying industry up until the present day: In the 19th century legislation, relating to planning, did not exist in the quarrying industry; this meant that even the small family quarries were not covered by appropriate laws. People had to rely on their own devices. My research has shown that no Planning Acts were instituted until 1947. Since that date legislation has been established to cover all aspects of quarrying, planning, as well as health and safety and the care of the environment. Legislation continues to be improved to make quarrying more environmentally friendly and to cover people's concerns. Legislation was established to try to address health and safety concerns, prompted by the increasing amount of accidents in mines and quarries in the 19th century (Mills, 2010 and Grahame French, Per. Comm. 16/08/2012). For example, quarrying for minerals such as limestone was considered to be the same as mining for coal. However, there are no similarities between coal mining, which is mostly conducted underground, whereas most limestone quarrying is done at surface level. In Shropshire, The Lilleshall Limestone Company was the only company to combine mining with their surface quarrying business. The consequences of limestone mining in Lilleshall are visible with subsidences around the area of the former Colliers Side Quarry; the most recent being in the 1980's. Examples drawn from records of mining for limestone in the 17th century in the Lilleshall area by the Leveson family, would indicate the lack of legislation covering either mineral mining, or health and safety issues.

In the 19th century "quarrying" was known as "mining". In the 1890s quarrying was covered by the Factory Acts. In a legal case of the 1890s, between Kent and Astley, it was argued that quarries were open to the air and thus should not be considered as a "factory". In 1851, a Royal Commission was initiated, chaired by Lord Astley, to investigate working conditions in the mining industry, which led to the first Inspector of Mines being appointed. Further, another Royal Commission, some ten years later, was set up to review working conditions in The Metalliaferous Mines Sector "quarries". This would lead eventually to The Metalliaferous Mines Regulations Act of 1872. However, the 1891 Factory Act altered the law to cover open air premises, "Quarries". In 1838, after a major accident in which 26 children died, Queen Victoria ordered an inquiry into safety issues in mining (Gallop, 2010). The quarry owners had many ongoing disputes about the exact terminology relating to surface quarries and so did not accept the 1878 Factory Act, maintaining that it did not apply to them. In 1892 the Secretary of State designated quarries as a "dangerous place of work", and therefore subject to special rules for quarries. This was followed two years later by the 1894 Quarries Act which was established to cover all surface mineral extraction. However, before this there had been The Quarry Fencing Act 1887, which was primarily to protect the general public from open land "quarries".

In the early 20th century, The Metalliaferous Mines Regulations Act was subject to a lengthy enquiry by a Royal Commission. The commission examined the general health and safety measures, as well as how to control dust. The 1910 Royal Commission also suggested the appointment of additional inspectors of mines. Other recommendations of the commission were that the mines and quarry proprietors should have their legal responsibilities clearly defined. It went further and strongly suggested that if there were twelve people working underground in mines, or in quarries, a registered manager should be in control (Mills, 2010, pp.39-69). The Royal Commission made many more recommendations that would benefit the mining and quarrying industries. The Mine Accident Act of 1910 was revised, suggesting that if ten people were employed, a first aid post should be set up on site. This was extended if there was an increase in the number of people employed on a site.

The Town and Country Planning Act of 1947 initiated a new planning system, which covered the ordinance of quarries and mines. Before the 1947 law, there was no planning control in the country, this new Act would prohibit any further kind of development whatsoever before, or after 1947. Quarrying operations that were taking place in Shropshire were covered by this new law and needed to apply for an Interim Development Order Consent (IDO) (Grahame French, Per Comm. 16/08/2012).

Throughout the 20th century, until the present day, planning legislation would continue to develop in the quarrying and mining industries. In 1951 an Act of Parliament was established after the Kennet Committee Report of 1938 and the Water Report of 1946. The 1951 Act would, for the first time, attempt to control environmental issues raised by the public concerning the quarrying and mining industries. The first recorded evidence I came across in my research concerning the effect of quarrying on the environment was at Wenlock Edge in the 1880s regarding the effect of quarrying on the environment, which was raised by the local community. These concerns would escalate in the late 19th and 20th centuries. After the 1947 Town and Country Planning Act, the updating of legislation applicable to the quarrying industry, became more demanding and would continue to improve substantially up until the present day.

With regards to public and employee safety, one would see the Fencing Act of 1887 upgraded by the Mines and Quarries Act, 1954. This act would go further, requiring the owners of the quarries to improve public safety; it also required proper health and safety and welfare for the employees. The 1969 Act, which covered Mines and Quarries (Tip), responded to public concerns raised in the aftermath of the Aberfan disaster but also had implications for disused mines and quarries. This was followed up in 1971 by the Mines and Quarries (Tips) Regulations, and the Quarries Regulations of 1999, which would require any instability, or movement, to be recorded and safety measures undertaken to correct them (The Mines and Quarries (Tips) Regulations, 1971 and The Quarries Regulation, 1999 www.legislation.gov.uk, accessed 29/10/2012).

In the late 1980s and early 1990s Interim Development Order Consent (IDOs) were given without proper planning controls, so in 1995 the government published Minerals Planning Guidance notes 14. This policy would stop any inconsistences in the implementation of mineral planning by different authorities. Throughout the early part of the 21st century the government would review and bring out different local government, planning and mineral guides, which now took into account the local environment and the impact quarrying was having on the countryside. The National Planning Policy Framework (NPPF) published in 2012 would see this policy simplify, streamline and replace all previous policies in the quarrying industry. The NPPF had new information contained in it particularly concerning "Conservation and Enhancing the Natural Environment and Conserving and enhancing the Historical Environment (Department for Communities and Local Government, National planning Policy Framework, March 2012, www.comunities.gov.uk, accessed 24/08/2012). This was a major development in government legislation in the quarrying industry requiring quarrying companies to undertake the recycling of stone and mineral wastes. Legislation is a major part of my research from the 19th century to the present day.

5. To record the technological developments on quarry sites i.e. tramways, inclines: The foundations for bulk mineral extraction had been laid with the purchase of former monastic lands after the Dissolution of the Monasteries in 1536-1541. The Lilleshall Abbey lands, as already mentioned, were purchased in December 1539 by James Leveson. In 1746, Sir John Leveson became The Earl Gower. Following this in 1764, the Leveson family reformed their business as The Lilleshall Partnership; he then created The Earl Gower and Company when he went into partnership with the Gilbert brothers. John Gilbert, was a prominent engineer and Thomas Gilbert, a Member of Parliament and legal adviser to the Company. The Lilleshall Company started construction of the Donnington Wood Canal in 1765 which was completed by 1767. The engineering design had been undertaken by John Gilbert. This was the first canal constructed in the county of Shropshire and would enable the Lilleshall company to transport its quarried limestone more economically.

At the end of the 1700's, limestone quarrying and mining was well established in the Lilleshall area. Engines were installed to pump water out of the mines, but this did not prove very successful. Earl Gower's quarries and mines were connected by branch lines to the main canal system, via Willmore Bridge. The canal wharf and bridge can still be seen today and gives one an insight into the advanced technology of the time. In the 1840s the railway was introduced and this made transporting processed limestone much easier.

Grinshill Stone Quarries Ltd in the late 19th century, introduced stone dressing machines powered by gas. This was a great innovation in processing stone block and avoided the blocks of stone splitting. Horse-drawn carts would transport the stone to the railway station; the stone then travelled onwards by rail. At the quarry site were cranes to lift the quarried sandstone blocks. On site visits I was able to see features of previous workings and foundations for cranes at the former Bridge, or Cureton, Stone Quarries.

However, one must remember that the geological formations of the landscape were different at every quarry and mine site. The River Severn played a large part in the transport of quarried minerals from across Shropshire, whereas minerals from Bayston and Brown Clee Hills, because of the awkwardness of their geographical location, relied on horse-drawn carts. Eventually the railways became pre-eminent. Today, at the Bayston Hill Quarries, the railway sidings are being reopened to transport quarried greywacke. The height of production at Wenlock Edge was in the 19th century; this would bring with it technological development in machines, inclines, tramways and railways. Lea Quarries, Wenlock Edge, from originally being family owned and operated in the 1840s to the middle of the 1950s, when it changed hands and was operated by the Ridgeway Limestone Ltd. This firm would install modern machines, such as primary processing, secondary crushing and screening plants; this is just a small example of the technological development at this site.

Llanymynech limestone quarries are unique, originally consisting of separately owned English and Welsh quarry. The limestone quarries at both sites had tramways and incline to transport the quarried limestone down the hill to their individual canals. Technological developments on the hill included Drum houses and a Tally House. Llanymynech had many lime kilns. Near the end of the 19th century The Warren Hoffmann Kiln was constructed to try to revive and save the limestone works. Although, the Warren Hoffmann Kiln was a leading technological advancement for its time and processed more quarried limestone at one time, yet it would not save the Llanymynech Limestone works.

Bayston Hill Greywacke Quarries are a 21st century site where modern machineries can be seen processing the quarried mineral for road construction. It also shows a quarry site working to protect the natural environment, which is now a required part of the planning process. By examining Bayston Hill and Grinshill Quarries one can see and understand what might have been happening in other redundant quarries across Shropshire.

6. To follow the innovations made in transport of quarried minerals by pack-horse, canal, railway and lorry: The original transportation of quarried minerals was done by pack-horse, or on the backs of people. For a long time the River Severn was used to transport quarried, or mined minerals, yet in some cases, access to this major river system of Shropshire was not possible because of the inaccessibility of some aspects of the geology of the landscape. This is where people and pack-horses were essential to the movement of quarried minerals down the hills, especially in the case of Brown Clee Hill.

Innovations in Shropshire's transport systems firstly came about with the construction of the canals in the middle of the 18th century. At this time, to access the new canal system, quarrying companies created canal wharfs where they could load the processed minerals onto barges waiting at the canal side. However, transporting lime by water was a dangerous process for the workers and processed lime could burn holes in the barges. Even so, it was a more direct and economic way to move the quarried minerals. Examples of canal wharfs can been seen at Llanymynech Limestone Works, where there are two canal wharfs, an English and a Welsh one, from where burnt lime could be taken down the canal, and local farmers could access limestone for fertiliser for their land.

After the coming of the canal system, the next transport innovation was the railway, which would make transporting quarried minerals much easier and able to reach more parts of the country. The new railway system met with opposition from major landowners, who at first were reluctant to allow the building of railway lines across their land. In order to subvert this problem, in some cases, the railway companies made use of earlier canal routes. In the early 19th century George Stephenson built a short-distance railway in Sunderland. However, by the middle of the 19th century railway companies created a countrywide network; this was the beginning of the decline of the canals system, which had revolutionised the transport system used by the quarrying industry. With the coming of the railway the quarrying industry of Shropshire expanded. The development of the railways would benefit the quarrying industry by making it cheaper to transport quarried minerals across Britain. Shropshire's 19th century quarrying industry expanded and continued to grow throughout the century. The quarrying companies built railway links from their quarries to connect to the main railway lines. This can be seen at Bayston Hill Greywacke Quarries and Brown Clee Hill.

The initial use of tracks and roads by horse and cart, was followed by canals and then railways to transport minerals. In the late 19th century and early 20th century at Brown Clee Hills gas powered lorries were developed. As lorries took over the greater part of transporting minerals, better road networks were needed. Road networks would be the next innovation in the transport system and would make it easier for quarry companies, where it had previously been difficult to connect to the railway. Raising finance for the development of new transport systems in the past had been gained through the issuing of shares. Canals and railways had been allowed to issue shares.

However, this was not the case with roads, so another way of raising money needed to be sought. In the 1700's, tolls raised for the construction and maintenance of roads was paid to local entrepreneurs who in turn were responsible for the upkeep of the turnpike roads; the toll paid was graduated in accordance with the goods being transported. Today, Bayston Hill Greywacke Quarries, because of its location, uses the A49 which in turn leads to the motorway system. Brown Clee Hill, because of it geological location, transferred its original use of the Ditton Priors Light Railway to road transport along the B4364, the Bridgnorth to Ludlow Road. Wenlock Edge quarries used the B4371 Church Stretton Road, an unsuitable narrow lane, which eventually caused road obstructions, as this local country road was originally designed for farm carts pulled by horses (Williams, 1997). Local people of Much Wenlock started to complain about the noise and heavy industrial traffic. This would eventually become one of the points that would be taken into consideration in future planning applications. Tarmac Ltd, at Bayston Hill, began to examine other options of transport, deciding to return to the railway by reopening their railway sidings, so linking the quarry to the main Shrewsbury to Cardiff rail line. The re-use of the railway system has enabled the return to a more environmentally friendly transport system. These innovations in the transport system transformed the quarrying industry of Shropshire.

7. How changes to quarrying were established to protected archaeological finds, the environment, ecology, restoration of the flora and fauna: Archaeology, whether beneath or above ground, includes anything that provides evidence of the impact of past human activities upon the landscape. Archaeological remains are important for both the archaeological, historical and environmental records. Shropshire's quarrying industry, from individuals, to families, and to large companies, gave precedence to the quarrying for minerals, over the need to record and protects archaeological finds. In the middle of the 18th century on Brown Clee Hill, hillfort ramparts were removed by quarrying, causing irreversible damage. Today one can only imagine what the complete hillfort must have been like and what it could have told the archaeological community. In 1907 to 1908 the Congress of Archaeological Societies, requested that all quarrying operation be suspended by the Brown Clee Hill Stone Company so that an archaeological evaluation could take place; this was refused. There was no legislation at the time to legally require the Brown Clee Hill Stone Company to comply with the Congress of Archaeology Societies' request. Today this does not happen because legislation protects what remains of the hillfort on Brown Clee Hill and any other archaeological finds beneath or above ground, but the effects of quarrying are visible on the hillfort and its landscape (Malim, *et al, 2011).*

Today in the 21st century, legislation requires that archaeological evaluations take place before any expansion of quarrying operation can be considered. An example of this requirement can be seen at Bayston Hill. In 2009, Tarmac Ltd of Bayston Hill, wanted to expand to the east of the present greywacke quarry. However, there was evidence of a known Roman road that crossed Bayston Hill from the south-east. So, Shropshire County Council, Planning Committee required that an archaeological investigation be carried out. This was conditional before consent could be granted for the expansion of quarrying operations (Malim, *et al* Hayes 2011). During the 1960s other sites of archaeological importance from the Neolithic and Bronze Age periods were discovered on, or around Bayston Hill.

The Environmental impact of the quarrying industry on the landscape has become of greater concern to the general public and people living near to quarrying operations. This was seen at Wenlock Edge in the 1880s when concerns were raised about the blasting times from one of the local quarries. Today any application for a quarry extension, by a quarry proprietor or operator, must include an environmental impact statement, which has to contain comprehensive information about the key environmental features which may suffer the effects of their proposed quarrying operations (Grahame French, per. comm. 196/08/2012).

When a proprietor or operator, submits a periodical review they are required to take into account ecological considerations, especially for Sites of Special Scientific Interests (SSSI). The government provides guidance notes on periodical reviews in NPPF, 2012; it also provides directions on the protection of the countryside, heritage and wildlife. The protection of the natural environment is a major issue today when planning application is considered by the Shropshire County Council, Planning Committee. Greater consideration is now given to the protection and conservation of the ecosystem, geology and valuable landscape. Other points taken into consideration by the Planning Committee now are: the effects of noise, of water pollution, from mineral waste both in the air and soil. These are just a few points raised. Within this book there are many more points which concern the safety of both the public and those working in the quarrying industry, but are primarily for the conservation and preservation of the natural landscape. The former Llanymynech quarries are a prime example of what happens to a redundant site. Today the former English and Welsh Quarries to the south and east of Llanymynech Hill are now known as the "Llanymynech Rock Wildlife Reserve". The importance of this reserve is the grassland flora and fauna contained within the site after nature has been allowed to reclaim the former quarries. This is an example of what has happened at just one former quarry site; other active sites, not only redundant quarries, that I have researched, have also important flora and fauna which are now protected by legislation (The Wildlife Trusts, www.shropshirewildlifetrust, accessed 18/10/2013).

The Department for Environment, Food and Rural Affairs, requires a high standard of restoration to be carried out at any mineral site, whether redundant, or before further applications for an existent active site, are applied. In the case of Bayston Hill Quarries operated by Tarmac Ltd, extraction of greywacke will cease in 2060. Tarmac Ltd will need to recreate grasslands, woodlands habitats and shrubs to be restored as areas of biodiversity. In concordance with this, work has already been undertaken to plant trees and shrubs on both sides of the public right of way. This is a prime example of what the government and local authorities now require of any quarry companies. They must restore the land that has been quarried to its former identity, or even better, keeping in mind the unique flora and fauna, so reclaiming all former quarry sites (The Department for Environment, Food and Rural Affairs, www.defra.gov.uk, accessed 22/07/2011).

8. The present-day approach to the restoration and legal requirements of after-care of redundant quarry sites: The Mineral Planning Authorities today require that a plan of action in writing by quarry companies be submitted to the local planning authority within three months after the restoration has been completed. The Planning Authority will need to approve the aftercare scheme and they will specify conditions required to restore the former quarry sites back to necessary standards for its subsequent use. The Planning Authority stipulates that former quarry sites should be restored, as far as possible, to their former agricultural use. In the case of the Bayston Hill Quarries, the topsoil and subsoil should be saved and, after the site is no longer required for quarrying, these should be reinstated so as to return the land to agricultural use. Woodlands should be restored to conditions capable of sustaining trees, shrubs, grassland areas and water features. The quarry face must be stabilised so that trees and other plants can be planted, thus making the area suitable for recolonization by flora and fauna.

The Mineral Planning Authority require that at the end of each year a written statement should be submitted outlining what work has been done and what future work will be carried out in the following year. In addition, they require that once the former quarry land has been fully restored, a minimum five year period is required for aftercare of the agricultural land, and a maximum twenty-five year period of aftercare is stipulated for woodlands, quarry benches, quarry faces, water features and habitat spaces (Shropshire Town and Country Planning Act 1990, accessed 21/07/2011 and Shropshire County Council Planning Committee, Proposals for Planned Restoration, accessed 16/03/2011).

7.1 Discussion

While major points of this PhD concern important industrial developments in limestone quarrying and mining of the 19th and 20th centuries, it has also researched into earlier archaeological and historical records within the county of Shropshire. This was done to provide a complete understanding of the importance of quarrying in Shropshire. It was interesting to understand how individuals, or families, owned and operated small quarries. The wills of these people show how important the small owned quarries were and how they were eventually distributed among family members.

The Dissolution of the Monasteries in the 1530s saw vast lands being purchased by prominent people. They knew of the mineral wealth beneath the former monastic lands and the money that could be made from extracting this resource. Before this time, it is shown that local tribes and then later the Romans, exploited these mineral resources in particular at Llanymynech Hill. The former Lilleshall Abbey area was the first place where innovations, such as the Donnington Wood Canal, were constructed along with other developments. Having the Gilbert brothers working for the Lilleshall Company was ideal; they contributed their experience and knowledge of engineering and the law. Other groups of people saw and grasped the opportunity to develop former monastic lands at Wenlock

Edge. The development of all the quarry sites that I have researched, was undertaken by entrepreneurs who saw the potential of the sites and the different minerals beneath the lands.

Mining for limestone in Shropshire was only carried out at Lilleshall. However, mining at this site struggled to succeed and eventually became unviable because of the danger from water flooding in, which brought about the closure of the mine. The coming of the canals in the late 18th century and then the railways and lorry transport would see Shropshire's quarrying industry develop and expand throughout the 19th and 20th centuries. The quarry companies constructed tramways which took the quarried minerals down to the new canal wharfs so that the extracted minerals could be transported along the canals. The coming of the railways would reach even more remote places and would create a network across Shropshire and the country. Although the major landowners were against the railways going across their lands at first, in time the benefit would soon be clear to all. The quarried minerals of Shropshire could be transported by rail more easily and further for less, so increasing the economy of both the quarry companies and the county. Road transportation with lorries was the final innovation and this was more advantageous for some quarry companies whose geological location made approach by canal or railway unreachable, as seen at Brown Clee Hill. However as with everything, there was a resurgence of the railway, and Bayston Hill Greywacke Quarries could see the benefit which could be obtained by reopening their railway sidings and linking back into the main rail network.

Legislation in early quarrying and mining did not exist until The Town and Country Planning Act of 1947. This book found that there were concerns about health and safety with regard to the accidents in mines and quarries, so that a Royal Commission was established to investigate and bring their findings to the government. In time, an Inspector of Mines was established by the government. In the 1880s one can see the Factory Acts were being applied to cover mining and quarrying, but prominent people like Kent and Astley argued that the Factory Acts were not applicable to open spaces. The 1890s would see the government bring into law the first Quarry Act. The Royal Commission of 1910 published its report over the years 1912 and 1914, which made extensive recommendations. They suggested that additional Inspectors should be appointed and there was an urgent need for the consolidation and simplification of the law.

From 1947 to the present day would see numerous planning laws being brought out and constantly re-examined and brought up to date with new parts and operation notes being added by the government. Even in the

1880s the public were concerned about blasting operations and the effect on the environment. From 1947 onwards quarrying operations and their ecological environmental effect have been a concern of the general public, especially when people lived very close to quarry sites. The NPPF published in 2012 is the most up to date legislation covering quarrying and the environment, ecology and flora and fauna. Today regulations leading to permission for planning consent for quarrying operations are very stringent and quarry companies have to submit written statements to the local planning authorities. These statements must cover how the quarry companies will address environment, ecology, flora and fauna, before they obtain consent to commence quarrying operations. Also, now when quarry companies want to move to another site, before this is permitted they must show in a written statement what restoration and after care they will put in place for the redundant quarry site. As one can see, there is a time during which a quarry company must look after a redundant site; in the case of Bayston Hill there will need to be 25 years of aftercare.

One can see from the research that the working life of the quarry workers was hard, dangerous and the working days long. In practice the quarry workers had to walk two hours from their houses to the quarry to start at 6 am, and then had to walk to the quarry face itself before starting their ten hour working day, before finally having to walk back home. Lilleshall and Brown Clee Hill quarry companies did build accommodation for their employees, however there were conditions to living in one of the cottages. The quarry workers would have to work longer hours in the summer months to cover for the winter months when working at the quarry face was not possible because of the poor weather. The quarry worker had no choice for if they did not work the longer hours in the summer, they and their families would have to leave the company cottages. Some quarry workers lived rough, sleeping in the open under hedges, this was dangerous and one could die from the cold.

Technology within the quarrying industry was constantly developing from new machineries such as the 19th century gas powered machines at Grinshill Quarries. In addition, when new companies took over Lea Quarries at Wenlock Edge, they upgraded the processing shed and other machines. At Llanymynech new technology can be seen with the construction of the Hoffmann Kiln in the late 19th century, but this would be too late to financially save the company. Today the former Llanymynech Lime Works Heritage Area was scheduled under the Ancient Monuments and Archaeology Areas Act 1979. This shows what can be done with a quarry site which has contributed to the industrial revolution of Shropshire and Britain.

Preservation and public awareness of quarrying remains low. When Wenlock Edge Limestone at Lea Quarry was closed and mothballed in 2007, there was potential to turn a unique survivor of the industry into an attraction. However, today all the machinery has been dismantled and taken away, the site being cleared for a business selling forestry products to earn revenue. This is a great shame because I suggest that it could have been made into a museum showing people of the 21st century how the quarried stone was processed by the different machines. When the Bayston Hill quarry is closed in the future it will hopefully be considered for preservation.

There are maps showing the quarrying evolution of Shropshire in the 19th and 20th centuries. They show the location of lime kilns and disused canals. The maps of Lilleshall provide details of the quarries, Barrack houses, lime kilns, canals and railway tracks. Brown Clee Hill maps also show the location of quarries, kilns and shafts; the last map of 1970s shows the landscape with no visible indication of the past industrial activities on the hill. However, if one were to undertake a field walk of the Brown Clee hill it would be possible to see the former quarry sites. Bayston Hill again shows the quarries and the location of Stone Age artefacts. Llanymynech Hill maps provide evidence of the quarries on the hill from 1880s to the 1950s. If one examines maps of Grinshill quarries, what is shown is a quarry inside the red boundary and one outside, but surrounded by Corbett Wood. Corbett Wood restricts the expansion of the quarry. Today there is only enough sandstone left for restoration work. These maps show how quarrying expanded and declined throughout the period researched.

Within the book there are many figures and tables showing the archaeological and historical records of the quarry sites researched in Shropshire. There are order and account books of transactions between Grinshill Quarries and different companies. An 18th century map of Bayston Hill, is suggestive of the location of the first quarry on the hill. There are many more maps and tables showing records of the quarry gangs, the amount of stone they quarried, and evidence of other activities including disputes between people, compiled by the quarry proprietors. This book writes about 19th and 20th century quarrying in Shropshire based on extensive research, it gives an extensive view of the importance of the mineral industry drawing on the archaeological and historical records of the county.

7.2 Conclusion

The word count limitation of the PhD has meant that I was restricted to what I could include in this book. Ironbridge Gorge Museum Trust holds an enormous amount of historical records of the activities of the Lillishall Company. The archives of Shropshire, Staffordshire and the National Library of Wales hold vast amounts of records of the quarry sites that I have researched. There is still documented evidence of the quarries that I could not use because of the word limit and there is still a lot of information for other people who would like to expand on my PhD. For example, I spent six weeks in the Ironbridge Gorge Museum Trust researching The Lilleshall records and could have written a book alone on this company. The order and account books of the Grinshill Quarries hold enormous valuable information on the activities of the company; I only looked at and used a few pages. This was the same across the whole research, there is a tremendous about of evidence in different places that could still be expanded on.

In this research limitation was caused because some companies, such as Tarmac Ltd, would not talk, or let me see their company archives, or even access the quarry. Through my research on Bayston Hill Greywacke quarries, I was able to contact local Councillor Ted Clarke who was the Chair of the Liaison Committee. He was very accommodating and kindly lent me the complete parish records with the Bayston Hill Quarry planning applications. When interviewing Councillor Ted Clarke, he told me of a Public Right of Way that when straight through the quarry sites. This gave an excellent opportunity to take photographs of the working quarry face and the processing machineries. I was also able to see the re-routed Public Right of Way and the planting of trees and shrubs, which were planted to act as a screen.

At Grinshill quarry the ownership changed from the English Clay Ltd in 1998 when it was sold it to Mr John O'Hare. Today, he and his wife Mrs Kathryn O'Hare, are both directors of Grinshill Stone Quarries Ltd. Mr Trevor Kirkham gave me a guided tour of Grinshill Quarries, which was an excellent way of understanding the background of the quarries. I did approach Mr John O'Hare the present owner of Grinshill Stone quarries Ltd for an interview, but without success, so without Mr Trevor Kirkham's knowledge of the quarries, I would not have gained so much information of the site. These are both examples of the restrictions I came across in my research; however, it also shows that there are ways around any obstacle that is put in the way of any investigation. This is something that any future researcher should take note off; that there is a way around anything if one thinks about it.

When reflecting on my research, is has been enlightening to access the archives at The Ironbridge Gorge Museum Trust Library, read books and journals on the subject. Interviewing people with their enormous knowledge on their particular quarries and gaining an understanding of their concerns, such as Mrs Harriet Devlin, on Llanymynech Quarries. All the people interviewed had different points to offer and concerns; however, Mrs Harriet Devlin impressed me because of her work and commitment to bringing the individual groups in England and Wales together to work on the protection of Llanymynech. Today the former Llanymynech Lime Heritage Area Works is protected under the Ancient Monuments and Archaeology Areas Act 1979. The Friends of Llanymynech have been, and continue to be, vital in the upkeep of Llanymynech Lime Heritage Area Works. This includes the knowledge that they pass on to the general public about the former works, the quarry workers, men who loaded The Hoffmann Kiln, the dangers which were faced and the coming of the canals. All of the quarry sites I researched are important and have something to offer to the archaeological and historical records. However, what has been achieved at the former Llanymynech Lime Works is a prime example of what can be done in restoration and conservation terms to protect the site. In addition at the former Llanymynech Lime Works it is good to see how the flora and fauna in the environment of this SSSI have been protected by legislation. All this contributes to an excellent story of the former quarry site, which tells its story to the people of the 21st century.

Other points to consider in this research are the legislation and mineral planning. It was interesting to scrutinize the legislation, or the lack of it, before The Town and Country Planning Act of 1947. Following on from 1851 with the setting up of the Royal Commission into safety at work its Chairman, Lord Astley, was instrumental in his fight to get better laws for the health and safety of all factory workers. The 1890s Factory Acts were designed to cover safety in employment in what was stated: ‘a factory’. This Act was objected to by mine owners who argued that as an enclosed factory floor bore no resemblance to an open quarry site, these Acts were inapplicable to a quarry worker’s situation. I suggest that the area of legislation and planning policies would benefit from further research by another PhD candidate. The developing legislation that I have researched needs to be investigated further to see how it will continue to change over time in the 21st century. I can see from my research how the legislation has changed and developed from the 19th, 20th and early 21st centuries.

As early as the 1880s the public raised concerns about the noise of blasting from quarrying operations. Quarrying operations have changed drastically over the period from the 18th, 19th and 20th centuries and continues to change in the 21st century. So legislation during these centuries has also developed and been constantly upgraded to adapt to changes in the quarrying industry, most recently as expressed in the NPPF published in 2012. One can see from this research that there has been a rise in concern for our environment, its ecology, and the protection of flora and fauna by SSSIs. The quarry companies now have to provide written statements with regards to the environmental impact that further quarrying may cause during the site's working life, as well as giving an undertaking for the restoration of the SSSI once quarrying has ceased. These statements are submitted to the local planning authorities when applying for extension to quarrying operations. So, a future PhD candidate could research further into these points and look into what the future in the 21st century holds in legislation and planning for the quarrying industry in Shropshire.

In the days of the early quarrying industry people did not live so close to the quarries. However more recently, at sites such as Grinshill Stone Quarries Ltd, people have purchased houses near to the quarry. This throws up numerous problems including those of noise, dust and transporting the quarried minerals all of which need to be taken into consideration, in my research this has been done. The need for quarried minerals is still as important today as it was in the 19th and 20th centuries. There are restoration projects that may need stone from quarries at some point in the future to match the existing building or bridges for instance as in the case of the English and Welsh Bridges in Shrewsbury. Bayston Hill Greywacke Quarry is a 21st century quarry and has a life until 2060 when it as to close. Greywacke

mineral is an essential commodity for the construction of roads in this country and in the world and can only being found in certain places in Britain; Bayston Hill in Shropshire is one of those places. So, this is a case where the local village, parish council, Shropshire Planning Committee and Tarmac Ltd need to live and work together.

Finally, these are some of the major points that can be studied in any future research; my book has laid the foundation for next PhD student. I cannot put enough emphasis that all bodies need to look at how the 21st century quarrying industry and the legislation policies; for a sustainable quarrying industry and environment, ecology in Shropshire. Shropshire's mineral industry has been a major contributor to the county's agricultural past and the industrial revolution of Britain. Shropshire was and still is unique because of the great variety of different minerals that can be found in the county and nowhere else in such abundance.

This has been an excellent opportunity to research into the important mineral industry of Shropshire's industrial past and into how it developed throughout each succeeding century to the present day. It is good to see that archaeological discoveries are now protected by the law and that quarry companies have allowed watching briefs and archaeological evaluation of their sites. The environment and ecology are also now considered important at every quarry site whether redundant or working. Restoration and conservation are now a major part of the Shropshire, Mineral Planning Committee commitment and has to be part of the quarrying industry today. More people should take the opportunity to appreciate and understand the enormous part that the Shropshire mineral industry has played in both this county's and the country's financial development, which in microcosm can be seen in Shrewsbury's town buildings and bridges.

BIBLIOGRAPHY

Primary Sources

Galloway, R. Photographs and Notes taken on site visit to Llanymynech Lime Works Heritage Area, 04/04/2013.

Galloway, R. Photographs and Notes taken on site visit to Lilleshall, 22/08/2009. Galloway, R. Photographs and Notes taken on site visit to Wenlock Edge, 14/04/2010. Galloway, R. Photographs and Notes taken on site visit to Bayston Hill, 29/03/2011. Ironbridge Gorge Museum Trust, Lilleshall Collection, 1998.320 (DLIL/3) 1992 (Memorandum of Agreement); 1998.320 (DLIL/3) 210a and 210b; (DLIL/3) 556a; 1997, 2202/38; 1997.320 (DLIL/3) 2202/5; 1998. 320 (DLIL/3) 174,175 (Assignment of Shares); 1998. 320 (DLIL/3) 178 (Articles of Co-Partnership), 1998.320 (DLIL/3) 237, accessed 09/09/2009.

National Library of Wales, Chirk Castle Estate, documents no, F6046 (Plan of proposed tramway 1804); F8528 to F8573; V/6050 and V6061; 8545-8573; F10119; F8528-8573; F8553 and F8571; F8612; F6057 (Valuation letter). Accessed 05/06/2013.

Personal communications:

Devlin, H. Per. Comm. Interview 27/06/2013.

French, M. Per. Comm. Interview 16/08/2012.

Kirkham, T. Per. Comm and Site Visit, 08/10/2011.

Clarke, T. Per. comm. Interview 03/03/2011.

White, R. H. Per. Comm. 04/08/2011.

Zorn, J. Per. Comm. Interview 04/04/2013.

Shropshire Archives:

Clee Hill Granite Company, document no 632/1 and 632/3, 22/12/2012.

Carline and Linnel's Order Book, document no. 800 Box 20c, accessed 12/10/2011.

Dickens to Hill, document no. 112/1/2380, accessed 12/10/2012.

Dividing and Inclosing the Commons or Waste Lands in the Townships of Grinshill, Sansaw and Clive in the Parishes of Grinshill and Saint Mary, in Salop, accessed 12/01/2012.
Document no 5359/1, accessed 22/12/2012.
Document No, XLS17956, The Hill Forts of Shropshire.
Documents Concerning The Township of Bayston Hill.
Documents no, 1224, box 108 deeds and documents, 1681/42/16.
Documents Relating Manors and Estates, 1206/99.
Documents Relating to Distributors, 3898/Rg/I.
Documents Relating to Farley Common and Bradley, 1224/2/75, 87 and 122.
Document no OD 64 CF and OD 64 V. F. accessed 22/12/2012.
Documents Relating to Unauthorised Quarrying, 1224/2/36 and 1224/2/116.
Excavations of Llanymynech, document no, C: 161.
Grinshill Parish Council, Documents, accessed 12/10/2001.
Interviews with Mr Robert. Cecil Cornfield, document no's, F L 56 and C 56.
Invoices for 1923-1937 and other documents no. 6057/2/2 to 6057/2/13.
Kelly's Directory of Herefordshire and Shropshire. (1895). London.
Kelly's Directory of Shropshire, 1871, 1890, 1910, 1913, 1917, 1920, 1922, 1926, 1929, 1934, 1937, and 1941.
Landscape and Development Evaluation, document no, q FL 71.1.
Lilleshall Records 1998-320 (DLlL/3), accessed 07/08/2009.
Llanymynech Heritage Focus Group, Three-year Report 1998-2001, accessed 22/05/2013.
Llanymynech Parish, documents no, C: FL 64.
Map of Bayston Hill 1789 X552/4/15/1/6/1
Notes on the Clee Hill Basalts, document no 524/1, accessed 22/12/2012.
Papers Concerning Regulation of Quarries, document no, 552/18/9/28/1, accessed 22/12/20.

Porter's Directories, accessed 23/04/2010.

Quarries Act 1894 and Letter, document no, 465/285.
Ref: DA6/100/8, Wenlock Council Minute Book, 1953.
Ref; DA6/101/2, Much Wenlock Council Minutes October 1961.
Ref: DA/130, Much Wenlock, Town Clerk, accessed 23/04/2010.
Ref: DA6, Wenlock Rate Book, 1838, accessed 23/04/2010. Shropshire Archives, Ref: DA6/101, Much Wenlock Council Minute May 1948, accessed 23/04/2010.
Shropshire Chronicle, 12 July 1918, accessed 23/04/2010.

Staffordshire Record Office, Sutherland Collection, D593/1/1/32 and D593/M/17/1, accessed 10/11/2009).

Shropshire Magazine April 1965, accessed 05/08/2009).

Shropshire site and Monument Records. No 7158, accessed 07/08/2009.

Sir Andrew Corbett and Acton Reynald document no. 322/2/428/1, accessed 12/10/2011.

The Account Book of Samuel and George Cank's Business for 1870 to 1873, document no. 4756/1/22.

The Wellington Journal, June 1860; 1869, accessed 06/08/2009.

1846 Tithe map of Much Wenlock.

Wages ledger for 1923-1937 and other documents no. 6057/2/2 to 6057/2/13.

Shropshire County Council, accessed12/10/2011.

Bayston Hill Parish Plan Action Plan, September 2006.

Mineral Extraction on Wenlock Edge: discussion paper and Report of Survey. (Shropshire County Council, 1979).

Committee, 10th October MN/2001/0768/NS

Planning Committee 9th April 1996: MS1992/0841/SY.

Planning Committee, February 2001: MS2005/1165/SY.

Planning Committee, 20 March 2007: MS2003/0529/SY.

Planning Committee 4th November 2008, MN2008/1387/NS.

Planning Permission, Town and County Planning Act 1990, MS2003/0529/SY.

Structure Plan: Written Statement: Mineral Extraction on Wenlock Edge. (Shropshire County Council, 1978).

The Quarry, (1898). *The Clee Granite Company's Quarries.* (London).

Staffordshire Archive

Bradford Collection, document no, D1287/13/86 (K/264) Potter letters; document no, D1287/18/31 (K/164) Griffiths letters; document no, D1287/M/2246, Plan of Land Leased to Colonel West at Llanymynech; document no, D1287/18/31 (K/164).

Staffordshire Record Office, Sutherland Collection, D593/J/13/2/1, 13/10/1625; D593/M/17/5, D593/B/2/7/22; D593/B/2/7/22/1 and D593/B/2/7/22/2,

Secondary Literature

Adams, D. R. (1970), (Ed and updated) Pearce, A. J. (1992). *Account No. 14. The Mines of Llanymynech Hill SJ 262218* (Shropshire, Shropshire Caving and Mining Club).

Adams, D. (2007). *The History of Limestone Mining In Church Aston and Lilleshall.* (Gloucestershire, Sutton Publishing Limited).

Adams, M. S. and McManus, F. (1994). *Noise and Noise Law: A Practical Approach.* (Chichester, Chancery Law Publishing Ltd).

Albert, W. (1972). *The Turnpike Road System in England, 1663-1840.* (London, Cambridge University Press).

Alfrey, J and Clark, C. (1993). *The Landscape of Industry.* (London, Routledge).

Armstrong, W. (1984). *Limestone Mine in the Wrekin Area: Phrase 1 Report.* (Staffordshire, Newcastle).

Bagshaw, S. (1851). *History, Gazetteer and Directory of Shropshire.* (Sheffield, Samuel Bagshaw).

Baker, N. 2010, *Shrewsbury; An Archaeological Assessment of an English Border Town.* (Oxford, Oxbow Books).

Barratt, G, Marriott, J, Reid, M, and White, R.H, (2008) *A Conservation Plan for the Historic Mining and Quarrying Landscape of the Clee Hills, South Shropshire. "Vol. 1. and 2, Chronological Summary, Significance of the site and management Recommendations.* Available at: http://archaeologydataservice.ac.uk/archives/view/cleehill_eh_2009/overview.c fm

Black, J. and Macraild, D. M. (2003). *Nineteenth-Century Britain.* (Hampshire, Palgrave Macmillan).

Blackwall, A. 1985. *Historic Bridges of Shrewsbury.* (Shrewsbury, Shropshire Libraries)

Boughey, J. (1998). *Hadfields's British Canals: the Inland Waterways of Britain and Ireland.*

Bridge, D, Sumbler, M. G. and Shepley, M. G. (2001). *A Hydrogeological Characterisation of the Superficial Deposits of the Severn Valley upstream of Shrewsbury.* (British Geological Survey Commissioned Report CR/01/187).

British Geological Society and Department of Environment, Transport and the Regions. (1998) *Mineral Resource Information for Development Plans* (Norwich, Her Majesty's Stationary Office).

British Geological Survey. (1990). *Mid Wales and Marches: Wenlock Edge.* (Southampton, British Geological Survey, Natural Environment Research Council).

British Geological Society (1998) *Technical report WF/98/6.*

British Geological Survey (1998) *Mineral Resources Information for Development Plan.* (Southampton, British Geological Survey, Natural Environment Research Council).

Brown, I. J. (1966). *A History of limestone Mining in Shropshire.* (Shropshire, Moore Brown, I. J. (1976). *The Mines of Shropshire* (London, Moorland Publishing Company).

Brown, I. (2005). *A List of Fatal Accidents in Shropshire Mines from 1850-1979: with some details of the major incidents.* (Shropshire, Shropshire Caving Club).

Bryan, Di, (2006). *Ditton Priors: A Settlement of Brown Clee* (Herefordshire, Logaston Press).

Burton, A and Pratt, D. (2001). *The Anatomy of Canals; The Early Years.* (Gloucestershire, Tempus Publishing Ltd).

Chapman, N., Evans, R., Poyner, D., and Powell, S. (1995) 'Forest of Wyre and Clee Hills Coalfields' in Pearce, P (Ed). *Mining in Shropshire.*Phillimore. [page numbers]

Chisholm, H. (1911) (ed). *Encyclopaedia Britannica.* (Cambridge, Cambridge University Press).

Clifton-Taylor and A.S. Ireson, 1994, *English Stone Building.* (London, Cassell Group).

Cossey, P. J, Adams, A.E., Purnell, M.A., Whiteley, M. J., Whytre, M.A. and Wright, V. P. (2004). *Geological Conservation Review Series No 29. British Lower Carboniferous Stratigraphy.* (Peterbourgh, Joint Nature Conservation Committee).

Currie, C. R. (ed) (1998). *The Victoria History of Shropshire: Wenlock, upper Corve-Dale* Vol. X. (Oxford, Oxford University Press).

Downes, J. (1980). *Grinshill Stone Quarries, Shropshire: A Study of Past, Present and Future Use.* (University of Durham).

Fewtrell, J. (1877) 'Parochial History of Llanymynech.' *Collections Historical and Archaeological relating to Montgomeryshire and its Borders,* Vol X [page nos.]

Fewtrell, J. (1878) 'Parochial History of Llanymynech.' *Collections Historical and Archaeological relating to Montgomeryshire and its Borders,* Vol XI [page nos.]

Fewtrell, J. (1878) 'Parochial History of Llanymynech.' *Collections Historical and Archaeological relating to Montgomeryshire and its Borders,* Vol XII, pp. 109-166

Gaffney, V. L., White, R. H. and Goodchild, H. (2007). *Wroxeter, The Cornovii and the Urban Process: Final Report on The Wroxeter Hinterland Project 1994-1997" Vol, 1, Researching The Hinterland.* Journal Of Roman Archaeology Supplementary Series 68. (Portsmouth, Rhodes Island).

Gale, W.K.V., Nicholls, C. R. (1979). *The Lilleshall Company Limited a History 1764-1964.* (Derbyshire, Moorland Publishing Co. Ltd).

Gallop, A. (2010). *Victoria's Children of the Dark* (Gloucestershire, The History Press,).

Gee, E. A. (1937). *A Historical and Archaeological Account of Wenlock Priory and its dependent Churches from the Earliest times to 1307.* (Unpublished Thesis for Degree of MA., University of Birmingham).

Gough, R. (1635-1723). Hey, D. (Ed). (1981). *The History of Myddle* (Harmondsworth, Penguin).

Greenwell, A. and Elsden, J. V. (1913). *Practical Stone Quarrying.* (London, Crosby Lockwood and Son).

Hadfield, C. (1969). *The Canals of The West Midlands.* (Devon, David and Charles Ltd).

Hadfield, C. (1979). *British Canals.* (Devon, David and Charles Ltd). Halcrow, W. Partners (1997). *Lilleshall Limestone Mines –Zone 2 Stabilisation Project.* (unpublished).

Hankinson, R. and Jones, N. (2011) *Llanymynech Quarry Drum House Renovation Works: Archaeological watching Brief and Recording.* CPAT Report No 1070. for Friends of the Llanymynech Limeworks Heritage Area. (Powys, The Clwyn-Powys Archaeological Trust).

Hannaford, H.P. (1996) *A Watching Brief of Abbey Foregate, Shrewsbury.*

Shropshire: Archaeology Service. Report No 100. September 1996. (Shropshire County Council).

Hannaford, H. R and Hanna, T. (2008) *Archaeological Monitoring of Work at the Llanymynech Lime Works Heritage Area, Shropshire 2008.* Archaeology Service Report 265, Shropshire County Council, For the Llanymynech Lime Works Heritage Project. (Archaeology Services, Shropshire County Council).

Heap, D. (1947). *Introducing The Town and Country Planning Act, 1947.* (London, Law Society).

Hewitt, P.B. (1991). *The Mining, Quarrying and allied Industries of The Clee Hill Region from the 1800s to 1930.* (Unpublished Thesis presented to the University of Birmingham).

Hogg, A. H. A. (1976). *Hill-forts of Britain.* (London, Hart-Davis MacGibbon). Holmes, D. A. (1987). *The Workings of the Silurian Wenlock Limestone in South-East Shropshire.* (u/published)

Jackson, D.A., Biek, L. and Dix, B. F. (1973) 'A Roman Lime Kiln at Weekley, Northants' *Britannia* 4, pp. 128-140

Johnson, D. (2002). *Limestone Industries.* (Gloucestershire, Tempus Publishing Ltd). Jones, K. (1998). *The Wenlock Branch: Wellington to Craven Arms.* (Mon, The Oakwood Press).

Jones, N. W. (2004) *Llanymynech Heritage Area Archaeological Survey.* CPAT Report No 618. Report for Oswestry Borough Council. (Powys, The Clwyd-Powys Archaeological Trust).

Jones, N. W., Hankinson, R. and Silvester, R. J. (2012) *Llanymynech Hill: Cultural Heritage and Management.* CPAT Report No 1166. Report for Cadw. (Powys, The Clwyn-Powys Archaeological Trust).

Jones, N. W., Silvester, R. J. and Britnell, W. J. (2003) *Montgomery Canal Conservation Management Strategy: Landscape Archaeology Assessment.* CPAT Report no. 550. (Powys, The Clwyn-Powys Archaeological Trust).

Lapworth, C. and Watts, W. W. (1910). *The Geology of Shropshire" Vol 2. 18881913.* Geologists' Association. (London, Turner and Co).

Lester, D. (1981). *Quarrying and Rockbreaking: the Operation and Maintenance of Mobile Processing Plants.* (London, Intermediate technology Publication Ltd).

Lloyd, C.L. (1968), *Borough of Much Wenlock: Wenlock 1468-1968* (The Borough Council of Much Wenlock).

Malim, T and Hayes, L. 2011 'An Engineered Iron Age Road, Associated Roman Use (Margary Route 64), and Bronze Age Activity Recorded at Sharpstone Hill, 2009.' *Transactions of the Shropshire Archaeological and Historical Society* 85, pp.7-80.

Mills, C. (2010). *Regulating Health and Safety in the British Mining Industries, 1800-1914.* (Surrey, Ashgate Publishing Limited).

Mingay, G. E. (1979). *Rural Life in Victorian England.* (London, Heinemann).

More, C. (1997). *The Industrial Age: Economy and Society in Britain 1750-1995.* (London, Longman).

Morriss, R. K. (1991). *Canals of Shropshire.* (Shrewsbury, Shropshire Books).

Morriss, R. K. (1991). *Railways of Shropshire.* (Shrewsbury, Shropshire Books).

Murchison, R. I. (1839). *The Silurian System.* (London, John Murray). Musson, A. E. (1978). *The Growth of British Industry.* (London, Batsford Ltd). Newman, J. and Pevsner, N. (2006). *The Buildings of England: Shropshire.* (London, Yale University Press).

Newman, P. (ed.) 2016 *The Archaeology of Mining and Quarrying in England A Research Framework for the Archaeology of the Extractive Industries in England Resource Assessment and Research Agenda* NAMHO Available at:
http://www.namho.org/documents/MINING_FRAMEWORK_Pt_one.pdf

Ove Arup and Partners. (1987). *Limestone Mines in the Wrekin Area: Lilleshall* Palmer, J. E. (1910). *British Canals Problems and Possibilities.* (London, T. Fisher Unwin).

Parker-Pearson, M., Bevins, R., Ixer, R., Pollard, J., Richards, C., Welham, K., Chan, B., Edinborough, K., Hasmilton, D., Macphail, R., Schlee, D., Schwenninger, J-L., Simmons, E. and Smith, M. (2015) 'Craig Rhos-y-felin: a Welsh bluestone megalith quarry for Stonehenge.' *Antiquity* 89 (348), pp1331-1352 Pearce, A, (1995) (ed), *Mining in Shropshire.* (Shropshire, Shropshire Books). Pinnell, J. (1999). *Wenlock Priory: Shropshire.* (London, English Heritage).

Pocock, R.W. and Whitehead,-T. H. (1948). *British Regional Geology: The Welsh Borderland.* (London, Her Majesty's Stationary Office).

Price, M. R. C. (1976). *The Cleobury Mortimer and Ditton Priors Light Railway.* (West Sussex, Oakwood Press).

Public General Acts of the United Kingdom Great Britain and Ireland. (1886) *Act of Parliament began and Holden on the 5th day of August 1886* (Streeten, F.E, Proprietor of the Law Journal Reports at No.5. Quarry Court, London).

Public General Acts of the United Kingdom Great Britain and Ireland. (1878) *Act of Parliament* (Streeten, F.E, Proprietor of the Law Journal Reports at No.5. Quarry Court, London).

The Quarry Managers' Journal, Vol 52. No 2. February 1968. (London).

The Quarry Managers' Journal' Vol 58, No. 3. May 1974. (London).

Ratcliffe, T. (2004) *Llanymynech Limeworks Heritage Area: Conservation Plan.* (Shropshire, Tim Ratcliffe Associates).

Rhodes, N. (2008). *The Story of Llanymynech and Pant.* (Llanymynech, Neil Rhodes Books).

Robbins, M. (1998) *The Railway Age* (Manchester, Mandolin, Manchester University Press.)

Scard, M. A. (1989) 'The Development and Changing Organisation of Shropshire's Quarrying Industry.' *Industrial Archaeology Review*. Vol. 11, No 2, pp. 171-186.

Sherlock, H and Pikes, P. J. (2001) *Llanymynech Heritage Area, Shropshire Archaeology Survey.* (Herefordshire, Archenfield Archaeology Ltd).

Simmons, I. G. (2003). *The Mooreland's of England and Wales: An Environmental History 8000BC-AD2000.* (Edinburgh, Edinburgh University Press).

Sivester, D. J., Owens, R. M. and Thomas, A. T. (1989). *Silurian Field Excursions.* (Cardiff, The Geologists' Association Cardiff).

Slater's Directory of Gloucester, Hereford, Monmouth, Shropshire and North and South Wales. (1859). (London, Isaac Slater).

Stanier, P. (2000). *Stone Quarry Landscapes: The industrial Archaeology of Quarrying.* (Gloucestershire, Tempus Publishing Ltd).

Thomson, D. B. (2004) *The Role of Grinshill Stone Quarries (1923) Ltd in the Rebuilding of The English Bridge in Shrewsbury.* Clive and Grinshill Conservation Group Occasional Publication No. 1. (Shrewsbury, Michael Lillistone).

Thompson, D. B. (2004) *The Grinshill Quarries from their Heyday in Victorian Times until 1923.* Clive and Grinshill Conservation Group occasional Publication No. 2. (Shrewsbury, Michael Lillistone).

Thompson, D. B. (2005) The Grinshill Sandstone Quarries in Relation To the Age of Improvement and the Second Period of national Building 1769 to 1840. "Clive and Grinshill Conservation Group occasional Publication No. 3. (Shrewsbury, Michael Lillistone).

Thompson, D. B. (2006) *Quarrying The Grinshill Stone for Buildings, Bridges and Roads 1540-1768.* Clive and Grinshill Conservation Group occasional Publication No. 4 (Shrewsbury, Michael Lillistone).

Thompson, D. B. and Lillistone, M. (2008) *The Decline and Resurrection of Quarrying at Grinshill 1927-1985*. Clive and Grinshill Conservation Group Occasional Publication No. 5. (Shrewsbury, Michael Lillistone).

Toghill, P. (2006). *Geology in Shropshire.* (2nd edn.) (Shrewsbury, Swan Hill Press). Toghill, P and Chell, K. (1984). *Shropshire Geology-Stratigraphic and Tectonic History Field Studies 6.* (Field Studies Council).

Topping, P. 2011a *Introductions to Heritage Assets: Pre Industrial Mines and Quarries* available at https://historicengland.org.uk/research/current/discover-andunderstand/industry-and-infrastructure/mining-and-mining-settlements/

Topping, P. 2011b *Pre Industrial Mines and Quarries* Historic England, available at: https://content.historicengland.org.uk/images-books/publications/ihapreindustrial-mines-quarries/preindustrialminesquarries.pdf/, accessed 17/10/2016.

Trinder, B. (2016) *The Industrial Archaeology of Shropshire* Logaston Press Trinder, B. S. (2000). *The Industrial Revolution in Shropshire.* (Chichester, Phillimore).

Tyler, R., White, R. and Galloway, R. S 'Lea Quarries, Much Wenlock, Shropshire, Archaeological Record and Heritage Assessment November 2009.' (Birmingham Archaeology).

Ward, T. O. (1841). *The Medical Topography of Shrewsbury and its Neighbourhood.* (Worcester, Deighton and Co).
Watson, J. (1911). *British and Foreign Building Stones.* (Cambridge, The University Press).
Watson, B. and Phillpots, C. (2015). 'The Old Welsh Bridge, Shrewsbury: Excavations at the Severn Theatre Venue: Frankwell, Shrewsbury, 2006-2007.' *Transactions of Shropshire Archaeology and Historical Society* 90, [page numbers]
Webster, G. (1993) *Rome Against Caratacus: The Roman Campaign in Britain.* (London, Routledge).
White, R. and Barker, P. (1998). *Wroxeter: Life and Death of a Roman City.* (Stroud; Tempus Publishing Ltd).
White, R.H. and Wigley, A. 2011 'Sharpstone Hill and Meole Brace in the Iron Age and Roman Periods' *Transactions of the Shropshire Archaeological and Historical Society* 85, 1-6.
Williams, J.G. (1997). *The Wenlock Limestone Industry.* (privately published by author).
Willies, L. (1994) *Mining before Powder.* Bulletin of the Peak District Mines Historical Society, Vol 12, No 3 (Historical Metallurgy Society Special Publication).
Wolmar, C. (2007). *Fire and Steam: How the Railways transformed Britain.* (London, Atlantic Books).
Yates, B. (2008). *The Railways and Locomotives of the Lilleshall Company.* (Bedfordshire, Irwell Press Ltd).

Web-Sites

Abdon, A History of the County of Shropshire, vol 10, wwwbritish-history.ac.uk, accessed 08/04/2013
Ancient Monuments and Archaeological Areas Act 1979, www.legislation.gov.uk, accessed 20/07/2012
Anglo-Saxon Shropshire, www.discovershropshire, accessed 27/03/2013.
Bat Conservation Trust, www.bat.org.uk, accessed 21/10/2013.
Bristar, www.degussa.cc.co.uk, accessed 10/04/2012.
British Aggregates Association, www.british-aggregates.co.uk/about/restoration accessed 22/07/2011.
British Geological Survey: Quarrying and The Environment, www.bgs.ac.uk/mendips/aggregates/environment, accessed 20/07/2011.
British Geological Survey: the Economic Importance of Minerals to the UK, www.mineralsUK.com, accessed 06/04/2011).

Campaign to Protect Rural England, Shropshire Branch, www.cpreshropshire.org.uk, accessed 06/04/2012.
Cardox International Limited, www.cardox.co.uk, accessed 10/04/2012.
Citizendia. http://citizendia.org/Bayston_Hill, accessed 28/04/2011.
Communities and Local Governments, www.communities.gov.uk/documents/pla, accessed 11/07/2011.
Degussal. www.procureit.com, accessed 28/03/2012.
Department for Communities and Local Government, "Planning and Mineral: Practice Guide" 2006. www.comunities.gov.uk, accessed 24/08/2012.
Department for Communities and Local Government, Environmental Impact Assessment and Reviews of Mineral Planning Permission, 2008, www.communities.gov.uk, accessed 31/01/2012.
Department for Communities and Local Government, Mineral Policy Statement 1 and 2, www.communitees.gov.uk, accessed 19/04/2012.
Department for Communities and Local Government, Minerals Planning Guidance 14, 1995, www.communities.gov.uk, accessed 30/07/2012.
Department for Communities and Local Government, Minerals Policy Statement 1, 2006 and 2, 2005, mwww.communities.gov.uk, accessed 15/05/2012. Department for Communities and Local Government, National planning Policy Framework, March 2012, www.comunities.gov.uk, accessed 24/08/2012, pp.1.14.25-29.30-36 & 58.
Department for Communities and Local Government, www.communities.gov.uk, accessed 12/10/2012.
Discover Shropshire. www.discovershropshire.org.uk, accessed 09/12/2011.
England Census 1881, www.search.ancestory.co.uk, accessed 14/09/2010.
Grinshill Sandstone, www.grinshillstonequarries.co.uk, accessed 29/03/2012.
Historical Directories. www.historicaldirectories.org accessed 03/03/2011.
Historical Directories. www.historicaldirectories.org accessed 12/10/2011
Historical Directories. www.historicaldirectories.org. accessed 03/06/2010.
https://historicengland.org.uk/research/current/discover-andunderstand/industry-and-infrastructure/mining-and-mining-settlements/Lobo Ecology-Species and Habitat, www.loboecology.co.uk/species, accessed 21/07/2011.
Mineral Extraction and Archaeology: A Practice Guide, 2006, www.englishheritage.org.uk, accessed, 11/07/2012.

Mineral Extraction and the Historic Environment. www.helm.org.uk. Accessed 02/12/2011.

MineralsUK, www.bgs.ac.uk/mineralsuk/planning.legislation, accessed 19/10/2012.

Mines and Quarries (Tips) Act, 1969, www.legislation.gov.uk, accessed 29/10/2012.

Mines and Quarries Act, 1954, www.legislation.gov.uk, accessed 29/01/2012.

Natural England, www.naturalengland.org.uk, accessed 26/10/2012.

Natural England, www.naturalengland.org.uk/regionwest_midland, accessed, 12/01/2012

Natural England. www.naturalengland.org.uk, accessed 16/03/2010. Quarries and Ancillary Activities, www.environ.iq/on/publications, accessed 17/04/2012.

Replica English Brake Drum, www.discovershropshire.org.uk, accessed 06/08/2013. Restoration Success at Tarmac Quarry,

www.tarmac.co.uk/aboutus/news/restorationsuccessattarmac, access 22/07/2011.

Ruthin Castle, www.denbigshirearchives, accessed 29/08/2013

Shrewsbury Railway Heritage Trust. www.shrewsburyht.org.uj/history, accessed 11/03/2013.

Shropshire County Council, www.shropshire.gov.uk/corbet-wood-management-plan, accessed 04/08/2011).

Shropshire Geology. www.shropshiregeology.org.uk, accessed 24/02/2011. Shropshire Hills-Natural Areas. www.naturalareas.naturalengland.org.uk accessed 14/03/2013.

Stable Block Restoration, www.LlanymynechLimestoneQuarries, accessed 27/09/2013.

Tally House, Llanymynech,www.discovershropshire.org.uk, accessed 06/08/2013. Tally House, www.Llanymynechlimestonequarries/4Ftons.tallyhouse, accessed 27/09/2013.

The Department for Environment, Food and Rural Affairs, www.defra.gov.uk, accessed 22/07/2011.

The Mines and Quarries (Tips) Regulations, 1971, www.legislation.gov.uk, accessed 29/10/2012.

The National Heritage of List for England, www.LlanymynectLimestonequarries/1021412, accessed 15/10/2013.

The National Trust, www.nationaltrust.org.uk/Wenlock-edge, accessed 16/09/2016. The Natural Environment and Rural Communities Act 2006,

www.naturalengland.org.uk/oruwork/conservation/biodiversity/protect and nama, accessed 21/07/2011.
The Quarries Regulation, 1999 www.legislation.gov.uk, accessed 29/10/2012. The Warren Hoffman Kiln Restoration, www.LlanymynechLimestoneQuarries, accessed 27/09/2013.
The Warren Hoffmann Kiln, Llanymynech,www.discovershropshire.org.uk, accessed 06/08/2013.
The Wildlife Trusts, www.motwt.co.uk , accessed 18/10/2013 Bat Conservation Trust, www.bat.org.uk, accessed 21/10/2013).
The Wildlife Trusts, www.shropshirewildlifetrust.org.uk, 18/10/2013.
Thomas Savin. www.Llanymynech.Llimestonequarries, accessed 09/06/2013
Tramways at Llanymynech, www.discovershropshire.org.uk, accessed 21/03/2013.
Tunnels under the A483 from Llanymynech www.discovershropshire.org.uk, accessed 20/06/2013.
United Kingdom Mines and Quarries Regulations in 1910. www.wikipedia.org/wiki, accessed 27/09/2012.
Valley of Stone, www.valleyofstone.org.uk, 23/08/2016).
Victoria County History of Shropshire: www.british-history.ac.uk accessed 26/03/2010.
Welsh and English Brake Drum Houses, www.discovershropshire.org.uk, accessed 20/06/2013.
Wenlock Edge. www.shropshiregeology.org.uk, accessed 10/01/2010.

Figures and Tables

Chapter 1:

Scard, (1989) Figure 1: *Map of Shropshire showing location of quarries studied and principal transportation infrastructure contemporary with quarries.* In: The Development and Changing Organisation of Shropshire's Quarrying Industry. *Industrial Archaeology Review.* Vol. 11, No 2.

Chapter 3:

University of Birmingham: Figure 3.1: *Geological Map of Grinshill.* [Online] Available from: Geology Roam.Digimap.edina.ac.uk/geologyroam. [Accessed: 24/04/2012].

University of Birmingham: Figure 3.2: *Geological Legend.* [Online] Available from: Geology Roam.Digimap.edina.ac.uk/geologyroam. [Accessed: 24/04/2012].

Figure 3.3: Atcham Bridge. [Online] www.bing.com/images.search?q=Atham+Bridge+shrewsbury. [Accessed: 23/08/2016].

Thompson, (2004) Figure 3.4: *The Grinshill Quarries from their heyday in Victorian times until 1923.* In: Clive and Grinshill Conservation Group occasional Publication. (Shrewsbury, Michael Lillistone).

Thompson, (2004) Figure 3.5: *The Grinshill Quarries from their heyday in Victorian times until 1923.* In: Clive and Grinshill Conservation Group occasional Publication. (Shrewsbury, Michael Lillistone).

Thompson, (2004) Figure 3.6: *The Grinshill Quarries from their heyday in Victorian times until 1923.* In: Clive and Grinshill Conservation Group occasional Publication. (Shrewsbury, Michael Lillistone).

Thompson, (2004) Figure 3.7: *The above picture above shows men working in the sheds possibly dressing stone; other men are working with a wooden crane which lifts stone blocks.* In: Clive and Grinshill Conservation Group occasional Publication. (Shrewsbury, Michael Lillistone).

Thompson, (2004) Figure 3.8: *The Grinshill Quarries from their heyday in Victorian times until 1923.* In: Clive and Grinshill Conservation Group occasional Publication. (Shrewsbury, Michael Lillistone).

University of Birmingham: Figure 3.9: *Grinshill Quarries 1890.* [Online] Available from: Geology Roam.Digimap.edina.ac.uk/geologyroam. [Accessed: 03/04/2012].

University of Birmingham: Figure 3.10: *Grinshill Quarries 1900.* [Online] Available from: Geology Roam.Digimap.edina.ac.uk/geologyroam. [Accessed: 03/04/2012].

University of Birmingham: Figure 3.11 *Grinshill Quarries 1920.* [Online] Available from: Geology Roam.Digimap.edina.ac.uk/geologyroam. [Accessed: 03/04/2012].

University of Birmingham: Figure 3.12 *Grinshill Quarries 1950.* [Online] Available from: Geology Roam.Digimap.edina.ac.uk/geologyroam. [Accessed: 03/04/2012].

University of Birmingham: Figure 3.13 *Grinshill Quarries 1970.* [Online] Available from: Geology Roam.Digimap.edina.ac.uk/geologyroam. [Accessed: 03/04/2012].

Figure 3.13 *Corbett Wood.* [Online]: Shropshire.gov.uk/Corbett-wood-management-plan [Accessed: 03/08/2011].

Chapter 4:

Shropshire Archives: Figure 4.1: *Llanymynech Heritage Area: Landscape and Development Evaluation, documents no, q FL 71.1*, [Accessed 22/05/2013]. University of Birmingham: Figure 4.2: *Geological map of Llanymynech Hill* [Online] Available from: Geology Roam.Digimap.edina.ac.uk/geologyroam. [Accessed: 12/06/2013].

University of Birmingham: Figure 4.3: *Geological Legend* [Online] Available from: Geology Roam.Digimap.edina.ac.uk/geologyroam. [Accessed: 12/06/2013].

University of Birmingham: Figure 4.3: *Llanymynech Quarries 1880* [Online] Available from: Geology Roam.Digimap.edina.ac.uk/geologyroam. [Accessed: 22/05/2014].

University of Birmingham: Figure 4.4: *Llanymynech Quarries 1890* [Online] Available from: Geology Roam.Digimap.edina.ac.uk/geologyroam. [Accessed: 22/05/2014].

University of Birmingham: Figure 4.5: *Llanymynech Quarries 1900* [Online] Available from: Geology Roam.Digimap.edina.ac.uk/geologyroam. [Accessed: 22/05/2014].

University of Birmingham: Figure 4.6: *Llanymynech Quarries 1950* [Online] Available from: Geology Roam.Digimap.edina.ac.uk/geologyroam. [Accessed: 22/05/2014].

Galloway: Figure 4.7: *Llanymynech Heritage Area.* Welsh and English Tunnels. [Taken: 04/04/2013].

Shropshire Archives: Figure 4.8: *Llanymynech Heritage Area,* Llanymynech Tramways and Wharf. [Clwyd-Powys Archaeological Trust-Projects, accessed 21/03/2013].

Shropshire Archives: Figure 4.9: *English Incline looking up from the base.* www.discovershropshire.org.uk, [Accessed 25/09/2013].

Shropshire Archives: Figure 4.10: *The English Drum House.* Llanymynech Limeworks Heritage Project. [Accessed 16/05/2013].

Shropshire Archives: Figure 4.11: *The English Drum House.* Llanymynech Limeworks Heritage Project. [Accessed 06/80/2013].

Shropshire Archives: Figure 4.12: *The Welsh Inclines looking up from the bottom.* Clwyd-Powys Archaeological Trust-Projects [Accessed 16/05/2013].

Staffordshire Archives: Figure 4.13: *1821map showing the coloured portions of land.* Bradford Collection [Accessed 16/04/2013].

Staffordshire Archives: Figure 4.14: *Potters letters.* Bradford Collection, document no, D1287/13/86 (K/264 [Accessed 16/04/2013].

Shropshire Archives: Figure 4.15: *Llanymynech Limeworks Heritage Area Conservation Plan.* Shropshire County Council [Accessed 08/12/2013].

Chapter 5:

British Geological Survey: Figure 5.1: Bayston Hill Quarry Map of Mid Wales and Marches. www.mineralsUK.com, [Accessed 14/11/1990].

University of Birmingham: Figure 5.2: *Geological map of Brown Clee Hill* [Online] Available from: Geology Roam.Digimap.edina.ac.uk/geologyroam [Accessed: 19/12/201424].

British Geological Survey: Figure 5.3: *Map of Mid Wales and Marches: Index showing colours and symbols* [Online] www.mineralsUK.com, *[Accessed* 14/11/1990].

University of Birmingham: Figure 5.4: *Geological Legend* [Online] Available from: Geology Roam.Digimap.edina.ac.uk/geologyroam. [Accessed: 19/12/2012]. Shropshire Archives: Figure 5.6: Map of Bayston Hill, Dated 1789 Showing a Stone Crop [Accessed 11/03/2011].

University of Birmingham: Figure 5.7: B*ayston Hill 1880* [Online] Available from: Geology Roam.Digimap.edina.ac.uk/geologyroam [Accessed: 04/04/2014].

University of Birmingham: Figure 5.8: B*ayston Hill 1890* [Online] Available from: Geology Roam.Digimap.edina.ac.uk/geologyroam [Accessed: 04/04/2014].

University of Birmingham: Figure 5.9: B*ayston Hill 1900* [Online] Available from: Geology Roam.Digimap.edina.ac.uk/geologyroam. [Accessed: 04/04/2014].

University of Birmingham: Figure 5.10: B*ayston Hill 1950* [Online] Available from: Geology Roam.Digimap.edina.ac.uk/geologyroam. [Accessed: 04/04/2014].

University of Birmingham: Figure 5.11: B*ayston Hill 1970* [Online] Available from: Geology Roam.Digimap.edina.ac.uk/geologyroam. [Accessed: 04/04/2014].

University of Birmingham: Figure 5.12: B*ayston Hill 1990* [Online] Available from: Geology Roam.Digimap.edina.ac.uk/geologyroam. [Accessed: 04/04/2014].

University of Birmingham: Figure 5.13: Brown Clee *Hill 1890* [Online] Available from: Geology Roam.Digimap.edina.ac.uk/geologyroam. [Accessed: 04/04/2014].

University of Birmingham: Figure 5.14: Brown Clee *Hill 1900* [Online] Available from: Geology Roam.Digimap.edina.ac.uk/geologyroam [Accessed: 19/12/2012].
University of Birmingham: Figure 5.15: Brown Clee *Hill 1950* [Online] Available from: Geology Roam.Digimap.edina.ac.uk/geologyroam [Accessed: 19/12/2012].
University of Birmingham: Figure 5.16: Brown Clee *Hill 1970* [Online] Available from: Geology Roam.Digimap.edina.ac.uk/geologyroam [Accessed: 19/12/2012].
Shropshire Archives: Figure 5.17: *Abdon Clee Hill plant and Quarry.* Document no PH/A/16 [Accessed 30/11/2012].
University of Birmingham: Figure 5.18: *Flood Quarry known as Crane Hole or Lake Kilowat, on Abdon Burf, Clee Hill, Shropshire* [Online] Available from: www.geolocation.ws, accessed 12/04/213 [Accessed: 19/12/2012].
University of Birmingham: Figure 5.19: *Frame of old quarry building on Abdon Burf, on Abdon Clee Hill, Shropshire* [Online] Available from: www.geolocation.ws, [Accessed: 12/04/2013].
University of Birmingham: Figure 5.20: *Stone Crushing Plant on Abdon Clee Hill, Shropshire* [Online] Available from: www.geolocation.ws, *[Accessed:* 12/04/2013].
University of Birmingham: Figure 5.21: *Stone Crushing Plant, on Abdon Clee Hill, Shropshire* [Online] Available from: www.geolocation.ws, *[Accessed:* 12/04/2013].

Chapter 6:

University of Birmingham: Figure 6.1: *Geological map of Lilleshall Hill* [Online] Available from: Geology Roam.Digimap.edina.ac.uk/geologyroam [Accessed: 24/04/2012].
University of Birmingham: Figure 6.2: *Geological map of Wenlock Edge* [Online] Available from: Geology Roam.Digimap.edina.ac.uk/geologyroam [Accessed: 24/04/2012].
University of Birmingham: Figure 6.3: *Geological Legend Lilleshall* [Online] Available from: Geology Roam.Digimap.edina.ac.uk/geologyroam [Accessed: 24/04/2012].
University of Birmingham: Figure 6.4: *Geological Legend Wenlock Edge* [Online] Available from: Geology Roam.Digimap.edina.ac.uk/geologyroam [Accessed: 24/04/2012].

University of Birmingham: Figure 6.5: *Lilleshall 1900* [Online] Available from: Geology Roam.Digimap.edina.ac.uk/geologyroam [Accessed: 25/03/2012].

University of Birmingham: Figure 6.6: *Lilleshall 1920* [Online] Available from: Geology Roam.Digimap.edina.ac.uk/geologyroam [Accessed: 25/03/2012].

University of Birmingham: Figure 6.7: *Lilleshall 1950* [Online] Available from: Geology Roam.Digimap.edina.ac.uk/geologyroam [Accessed: 25/03/2012].

University of Birmingham: Figure 6.8: *Lilleshall 1960* [Online] Available from: Geology Roam.Digimap.edina.ac.uk/geologyroam [Accessed: 25/03/2012].

University of Birmingham: Figure 6.9: *Lilleshall 1980* [Online] Available from: Geology Roam.Digimap.edina.ac.uk/geologyroam [Accessed: 25/03/2012].

University of Birmingham: Figure 6.9: *Lilleshall 1990* [Online] Available from: Geology Roam.Digimap.edina.ac.uk/geologyroam [Accessed: 25/03/2012].

University of Birmingham: Figure 6.11: *Wenlock Edge Quarries and Limekilns 1890* [Online] Available from: Geology Roam.Digimap.edina.ac.uk/geologyroam [Accessed: 19/03/2014].

University of Birmingham: Figure 6.12: *Wenlock Edge Quarries and Limekilns 1900* [Online] Available from: Geology Roam.Digimap.edina.ac.uk/geologyroam [Accessed: 19/03/2014].

University of Birmingham: Figure 6.13: *Wenlock Edge Quarries and Limekilns 1950* [Online] Available from: Geology Roam.Digimap.edina.ac.uk/geologyroam [Accessed: 19/03/2014].

University of Birmingham: Figure 6.14: *Wenlock Edge Quarries and Limekilns 1970* [Online] Available from: Geology Roam.Digimap.edina.ac.uk/geologyroam [Accessed: 19/03/2014].

University of Birmingham Map Library: Figure 6.15: *Lilleshall map showing location of quarries and mines* [Accessed: 09/05/2010].

Brown, (1976) Figure 6.16: *Lilleshall Company Barrack Houses.* London, Moorland Publishing Company.

Birmingham Archaeology, (2009), Figure 6.17: *Primary Crusher.* Lea Quarry, Wenlock Edge.

Birmingham Archaeology, (2009), Figure 6.18: *Primary Crusher.* Lea Quarry, Wenlock Edge.

Birmingham Archaeology, (2009), Figure 6.19: *Primary Crusher showing chain screen.* Lea Quarry, Wenlock Edge.

The Quarry Manager Journal, (1968), Figure 6.20*: Primary Crusher with intake hopper with chain screen.* Vol 52. No 2. February (London).

The Quarry Manager Journal, (1968), Figure 6.21*: High Intensity Screen handling scalping plant* Vol 52. No 2. February (London).

Birmingham Archaeology, (2009), Figure 6.22: *West Screen Conveyor.* Lea Quarry, Wenlock Edge.

Birmingham Archaeology, (2009), Figure 6.23: *View of Hoppers.* Lea Quarry, Wenlock Edge.

Birmingham Archaeology, (2009), Figure 6.24: *West Screen Conveyor.* Lea Quarry, Wenlock Edge.

Birmingham Archaeology, (2009), Figure 6.25: *View of Hoppers.* Lea Quarry, Wenlock Edge.

Tables

Chapter 3:

Shropshire Archives: Table 3.1: *Carline and Linnel's Order Book,* Document No. 800 box, 20c, [Accessed 12/10/2011].

Thompson, (2004): Table 3.2: *wages ledger for 1926-1929.* In: Clive and Grinshill Conservation Group occasional Publication. (Shrewsbury, Michael Lillistone). *And other documents no. 6057.*Shropshire Archives [Accessed 12/10/2011].

Thompson, (2004): Table 3.3: *wages ledger for 1926-1929.* In: Clive and Grinshill Conservation Group occasional Publication. (Shrewsbury, Michael Lillistone). *And other documents no. 6057.*Shropshire Archives [Accessed 12/10/2011].

Thompson, (2004): Table 3.4: *wages ledger for 1926-1929.* In: Clive and Grinshill Conservation Group occasional Publication. (Shrewsbury, Michael Lillistone). *And other documents no. 6057.*Shropshire Archives [Accessed 12/10/2011].

Thompson, (2004): Table 3.5: *wages ledger for 1926-1929.* In: Clive and Grinshill Conservation Group occasional Publication. (Shrewsbury, Michael Lillistone). *And other documents no. 6057.*Shropshire Archives [Accessed 12/10/2011].

Chapter 4:

Staffordshire Archives: Table 4.1: *Bradford Collection, document no, D1287/13/86 (K/264) Potter letters,* [Accesssed16/04/2013].

The National library of Wales: Table 4.2: *Chirk Castle, document no, F10119,* [Accessed 05/06/2013].
The National library of Wales: Table 4.3: *Chirk Castle, document no, F8528-8573,* [Accessed 05/06/2013].
The National library of Wales: Table 4.4: *Chirk Castle, document no, F8545-8573,* [Accessed 05/06/2013].
University of Birmingham: Table 4.5: Tally House Ledger for the 22nd October, 1910: [Online] Available from: www.Llanymynechlimestonequarries/4Ftons.tallyhouse [Accessed 27/09/2013].
University of Birmingham: Table 4.6: Tally House Ledger showing destination of quarried limestone: [Online] Available from: www.Llanymynechlimestonequarries/4Ftons.tallyhouse [Accessed 27/09/2013]. University of Birmingham: Table 4.7: 1-5 Showing a number of men working at Llanymynech Lime Works: [Online] Available from: www.Llanymynechlimestonequarries/4Ftons.tallyhouse [Accessed 27/09/2013].

Chapter 5:

Shropshire Archives: Table 5.1: *Kelly's Directory Shropshire, 1871, 1890, 1900 and 1920,* [Accessed 03/03/2011], *documents concerning the Township of Bayston Hill,* [Accessed 24/02/2011].[Online] *Historical Directories, www.historicaldirectories.org,* [Accessed 03/03/2011].
Shropshire Archives: Table 5.2: *The Clee Hill Granite Company, document no 632/1*, [Accessed 22/12/2012].
Shropshire Archives: Table 5.3: *Shropshire County Council, Planning Permission, Town and County Planning Act 1990,* [Accessed 21/07/2011], and Adams, et al (1994). *Noise and Noise Law: A Practical Approach.* (Chichester, Chancery Law Publishing Ltd).

Chapter 6:

Ironbridge Gorge Museum Trust: Table 6.1: *Lilleshall Collection, 1998.320, (DLIL/3) 174, 176 Assignment of Shares and 1998.320, (DLIL/3) 178 Articles of Co-Partnership* [Accessed 05/08/2009].
Shropshire Archives: Table 6.2: *Kelly's Directory of Shropshire, 1913, 1917, 1922, 1926, 1929, 1934, 1937 and 1941* [Accessed11/06/2010]. [Online] *Historical Directories, www.historicaldirectories.org* [Accessed 03/06/2010]. Ironbridge Gorge Museum Trust: Table 6.3:

Lilleshall Collection, [Accessed 05/08/2009]. Shropshire Archives: [Accessed10/11/2009].

APPENDICES

Appendix 1: Showing account book

Table 3.2: The Account Book of Samuel and George Cank's Business 1870 to 1873:

1880 to 1873.	The Grinshill Co to Samuel Cank.	Shillings	£:s:	d
April 30th and	Ridding in Quarries and Tilling Cranes.		40: 3:	0
	Samples to London.		5:	0
May 31st, 1870.	Ridding in Quarry.		33: 12:	5
	Samples to London.		1: 4:	9
June 30th, 1870.	29" 65, 5" Stone Got and Loaded.	4/-	49: 8:	4
	22" 23, 1" Haulage to Yorton.	4¾	13: 17:	10
	265, 8" [Ditto] to Hordret.	5/-	5: 10:	5
	73 Loads of Burs.	6/-	1: 16:	6

	12: 12 tons to ….. to Yorton.	2/6	1: 11: 6
	Working ……. Stone		4: 0
		[Total]	72: 8: 7
	£: s: d		
	By Cash 26: 13: 3		
	[Ditto] 5: 6: 10		
	: 8: 17: 2		
	: 15: 7		
	Bill for Stone 15: 17: 7		
	Balance due 14: 18: 9		
	Total 72: 8: 7		
July 30th, 1870.	Received off W. Townsend the sum of thirty-five pounds ……. Of work at Grinshill Quarries.		
August 13th, 1870.	Received off W. Townsend the sum of forty-one pounds 12 shillings and 6 pennies ……. On account of work done at Grinshill Quarries. [It is signed] Samuel Cank.		41: 12: 6

July 30th, 1870	Balance brought over.		14: 18: 9
	20" 78, 11" Stone Got and Loaded.	4/-	34: 12: 8
	369, 0. Haulage to Yorton.	1¾	2: 12: 9
	839, 0. [Ditto] Hadnet.	5/-	17: 9: 7
	354, 3. :Albrighton.	3/-	5: 3: 3
	117, 0. :Salop.	4½	2: 3: 10
	102 Loads of Burs.		2: 11: 10
	George Haffa Bill.		9: 2
	Credit By Cheque.	[Total]	80: 2: 0
	[Ditto]		35: 0: 0
	Bill for Stone.		41: 12: 6
			3: 9: 6
		[Total]	80: 2: 0
August 31st,1870	To balance Red Quarry.		3: 12: 6 ½
	3"12, 0". Stone Omitted.	4/-	5: 5: 6 ½

July 30th, 1870.	3"12, 0. Haulage to Yorton.	1¾	2: 5: 6
	202" 6, 10". Got and Loaded.	4/-	33: 15: 4
	579, 7". Haulage to Yorton.	1¾	4: 4: 5
	611, 10. [Ditto]Hadnet.	5/-	12: 14: 7
	311, 6. :Albrighton.	3½	4: 10: 8
	55, 9. :Salop.	4½	1: 0: 7:½
		[Total]	67: 7: 8
	……. Over paid for Burs. Paid Sept 10th 1870		2: 11: 8
	Stone over paid George Cank, £ 56: 9: 4, [signed over a stamp]		8: 7: 4
	Deducted	[Total]	56: 9: 4
September 30th, 1870.	25" 73, 11". Got and Loaded.	4/-	41: 17: 8
	13" 85, 2". Haulage to Yorton.	1¾	10: 1: 11
	87, 9". [Ditto] Albrighton.	3½	1: 5: 4
	276, 3. : Salop.	4½	5: 3: 6

	384, 11. : Hadnet.	5/-	8: 0: 0
		[Total]	66: 8: 5
	Sept. 24th Cheque on Acct		30: 0: 0
	Balance.		36: 8: 5
	On Acct of Ridding Bridge Quarry		25: 0: 0
		[Total]	61: 8: 5
October 31st, 1870.	16” 13, 9”. Got and Loaded.	4/-	26: 17: 8
	160, 7. Haulage to Yorton.	1¾	5: 10: 10
	297, 7. [Ditto]Hadneton.	5/-	6: 3: 9
	113, 9. : Albrighton.	3½	1: 12: 11
	189, 9. : Salop.	4½	3: 10: 10
		[Total]	43: 16: 0

October 31st, 1870, continued.	…….. from Newport		2: 10: 0
	[Ditto] Salop.		13: 4
	…….. for Carriage		1: 19: 6
	Cank, Smith		4: 16: 9
	[Ditto].		5: 10: 6
	Straw and Thatching Stack.		1: 3: 0
		[Total]	60: 9: 1
	Deduct Stone Bill. [Final Total] £40: 10: 0		7: 8: 1
		[Total]	53: 1: 0
			30: 0: 0
	Oct. 22nd, 1870 on Acct.		23: 1: 0
	Deduct from [Total].		16: 19: 0
	On Acct of Ridding in Bridge Quarry.	[Final Total]	40: 10: 0

November 30th, 1870.	72"1, 0 Got and Loaded.	4/-	12: 0: 4
	115, 3 Haulage to Albrighon.	2½	1: 13: 6
	51, 0 [Ditto] Hadnet.	5/-	1: 1: 3
	172, 6 : Salop.	4½	3: 4: 6
		[Total]	17: 19: 7
	.... Wellington Acct deduct		2: 3: 0
		[Total]	15: 16: 7
November 19th	On Account of Ridding Bridge Quarry.		14: 3: 7
	[Signed over stamp], George Cank £30: 0: 0	[Final Total]	30: 0: 0
December 31st, 1870.	32" 7, 3. Got and Loaded.	4/-	5: 9: 0
	84", 7. Haulage to Hadnet.	5/-	1: 15: 0
	20 Loads of Burs Loaded.		10: 0
		[Total]	7: 14: 0
Dec. 17th	On Account Ridding Bridge Quarry.		22: 6: 0
Dec. 31st	[Ditto]	[Final	20: 0: 0

		Total]	
	Dec. 31st, 1870. George Cank, [signed of stamp] £50: 0: 0		50: 0: 0
January 31st, 1871.	2" 11, 10 Got and Loaded.	4/-	4:7: 0
Dec. 17th, 1870	101, 4 Haulage to Yorton.	1¼	14: 9
	117, 0 [Ditto] Salop.	4½	2:3: 10
	74, 4 : :	4½	1:7: 9
	20 Loads of Burs Loaded.	5/-	10: 0
	20: 2: 0 tons Burs to Yorton.	2/6	2: 10: 3
	1680 Cube yards of Ridding.	10/-	70: 1: 0
	"Canks bill for haulage for Cranes.		15: 0: 0
	……… to Carriage for ………		4: 12: 4
		[Total]	101: 5: 11
	Canks bill for stone.		4: 18: 10
	Overcharge to Kinnerly		7: 6
		[Total]	5: 6: 4
		[Total]	95: 19: 7

January 31st,	By Cash		£:
1871. Sept 20th	s:	d	
Oct 21st	[Ditto]		25:
	0:	0	
Nov. 30Th			
	:		16:
Dec. 17th	19:	0	
:31st	:		14:
	3:	5	
Jan. 28th, 1871.			
	:		22:
	6:	0	
	:		20:
	0:	0	
			30:
	0:	0	
		[Total]	128:
	8:	5	
		Pay Sheet	101:
	5:	11	
		Carried forward	27:
	2:	6	

	George Cank [signed over stamp] £101: 5: 11			
February 31st, 1871	[Unreadable]			
March 31st, 1871.	22"21, 4. Got and Loaded.		4/-	37: 0: 4
	1409, 7. Haulage to Yorton.		1¾	10: 5: 5
	232, 0. [Ditto]		4½	4: 7: 0
	34: 5: 0 .Tons Burs to Yorton.		2/6	4: 5: 7
	32"00, 0 …………		[Total]	55: 18: 4
	dPrint First 6	£: s: 18:15:	4/-	53: 6: 8
			[Total]	109: 5: 0
	By Cash	50: 0: 0		
March 25th,	………..	1: 8: 0		
[Ditto]·	……………..	14: 16: 0		
:	Cheque	30: 0: 0		
April 8th	[Total]	114: 19: 6		

	[Deduct] 109: 5: 0		
	5: 14: 6		
	[Stamp crossed by money] £109: 5: 0 signed George Cank		
April 30th, 1871.	27"83, 7. Got and Loaded.	4/-	46: 7: 8
	22"98, 0. Haulage To Yorton.	1¾	16: 15: 1
	178, 3. [Ditto] Albrighton.	3½	2: 11: 11
	62, 7. :Salop.	4½	1: 3: 3
	30: 1: 0 Tons Burs to Yorton.	2/6	3: 15: 0
		[Total]	70: 12: 11
	200 yards Sub Shadding to Yorton.	2/10	28: 6: 8
	Removing Crane … for conduct.		2: 0: 0
April 30th , 1871	Ref Legh, Bill for Plinth and Capping.		3: 17: 4
		[Total]	104: 16: 11
	£: s: d		
	….. Forms. 5: 14:6 1871, May 20th		

	Stone Bill 7: 10: 0 George Cank,		
May 20th	Cheque 91: 12:5 £104: 16: 11		
	[Total] 104: 16: 11		
May 31st, 1871.	42”36, 5. Got and Loaded.	4/-	70: 12: 0
	37”28, 4. Haulage to Yorton.	1¾	27: 3: 8
	180, 1. [Ditto] Albrighton.	3½	2: 12: 6
	53, 4. :Salop.	4½	19: 10
	150, 7. ::	4/-	2: 10: 0
	Work for Rip and Road. £:s: d.		1: 0: 0
June 3rd	Stone Bill. 4:2: 0[stamp crossed] July 1st		
July 1st	On Acct Cheque 50: 0: 0[signed George Cank]		
	To Balance 50: 16: 0		
	[Total] 104: 18: 0		
June 30th, 1871	48”15, 3. Got and Loaded.	4/-	80: 5: 0
	32”33, 3. Haulage to Albrighton.	2½	23: 11: 5

	29, 10. [Ditto]Salop.	4½	7: 12: 0
	408, 2. ::	4/-	1: 5: 0
	637, 1 : Leaton.	4½	11: 18: 10
		[Total]	125: 1: 9
	212 Yards Sub. Shadding.	2/10	30: 0: 8
	30: 14: 0 Tons Burs to Yorton.	2/6	3: 19: 3
	£: s: d	[Total]	159: 1: 8
July 1st	On Acct. 49: 4: 0		
	Stone Bill. 3: 10: 8		
July 15th	On Acct. 100: 0: 0		
: 29th	[Ditto] 50: 0: 0		
	[Total] 202: 14: 8		
	159: 1: 8 1871, July 29Th		
	43: 13: 0 Received £159: 1: 8. George Cank.		

August 31st, 1871	33"26, 5. Got and Loaded.	4/-	55: 8: 8
	22" 23, 0. Haulage to Yorton.	1¼	16: 4: 8
	781, 6. [Ditto] Salop.	4½	14: 12: 10
	147, 7. : Leaton.	4½	2: 15: 1
	38, 2. : Black Birches.	1¾	5: 0
Sept 9th	£: s:d	[Total]	89: 6: 3
	Cheque 60: 0: 0 1871, Sept 23rd received		
Sept 23rd	Stone Bill 2: 7: 10 George Cank £89: 6: 3		
	Cheque 10: 0: 0		
	[Total] 102: 7: 10		
	[Deduct] 89: 6: 3		
	13: 1:7		
September 30th, 1871.	46"81, 0. Got and Loaded.	4/-	78: 0: 4
	39", 21, 10. Haulage to Yorton.		28: 11: 11
	263 [Ditto] Salop. [The rest was unreadable.]	4½	10: 1: 11

November 30th, 1871.	41”82, 5. Got and Loaded.	4/-	69: 14: 0
	32” 98, 2. Haulage to Yorton.	1¾	24: 0: 11
	170, 2. [Ditto] Leaton.	4½	3: 3: 9
	369, 9. : Salop.	4	6: 18: 4
Oct 31st	108, 0. Omitted	4/-	1: 16: 0
	108: 0. : Yorton.	1¾	15: 9
	£: s: d Received Dec 30th, Dec 1871	[Total]	106: 8: 9
Dec 9th	Cheque 50: 0: 0 £106: 8: 9		
Dec 30th			
	[Ditto] 56: 8: 9		
	[Total] 106: 8: 9		
December 30th, 1871.	26” 26, 5. Got and Loaded.	4/-	43: 15: 4
	1543, 9. Haulage to Yorton.	1¾	11: 5: 0
	304” 9. [Ditto] Salop.	4½	5: 14: 0
	409, 7. : Leaton.	4½	9: 3: 9

		[Total]	69:	18:	1
	Deduct over Mining 23 tons	4/-		7:	8
	[Ditto] Carriage 170 tons	4½	3:	11:	5
		[Total]	66:	6:	8
	Ridding Red Quarry.		26:	0:	0
	Sir. …. Corbet, Bill.		18:	14:	8
	Removing 5 Cranes.		15:	0:	0
December 30th, 1871.	Compiling Bill for do …….		3:	13:	6
	Maddock Shadding.		61:	14:	5
	[Ditto] Quarrying		10:	17:	0
	Ridding Bridge Quarry.		31:	4:	4
	[Ditto] Red Quarry.		3:	5:	0
	Bowder and Co Shadding.		42:	3:	9
	Powell and Co. :		12:	4:	0
	Ridding Two Leads ………		29:	12:	6

	[Ditto] Bridge Quarry.		6: 8: 0
	Building Rough Wall.		1: 10: 0
	Blacksmith Bill.		7: 15: 3: ½
Jan 31st, 1872.		[Total]	336: 9: 3: ½
	Pay Sheet.		89: 7: 3
[Ditto]	£:s: d	[Total]	425:16: 4: ½
	By Cash. 7: 8: 0[Deduct From Total]		11: 12: 0
		[Total]	414: 4: 4: ½
	Stone Bill. 4: 4: 0Received April 2nd , 1872		
	[Total]. 11: 12: 0George Cank.		
January 31st, 1872.	Got and Loaded.	4/-	53: 2: 4
	31" 84, 9. Haulage to Yorton.	1¾	10: 2: 10
	261, 5 [Ditto] Salop.	4½	4: 17: 10
	713, 5 : Leaton.	4½	13: 7: 4
	Working Chim, Cap and Base for …………		4: 8: 6

	146, 11 Got and Loaded.	4/-	2: 8: 8
	146, 11 Haulage to Yorton.	1½	1: 1: 3
		[Total]	89: 7: 3
Jan 31st	£: s: d		11: 12: 0
	By Cash 7: 8: 0	[Total]	77: 15 3
	Stone Bill 4: 4: 0 Received March 2nd 1872.		
	[Total] 11: 12: 0		
February 29th, 1872.	34"61, 7 Got and Loaded.	4/-	57: 13: 8
	28"02, 7. Haulage to Yorton.	1¾	20: 8: 7
	172,1 [Ditto] Salop	4½	3: 7: 1
		[Total]	81: 9: 4
	177, 6 Stone Omitted.	4/-	2: 19: 0
	117, 6 Haulage [Ditto].	1¾	17: 0
	60, 6 : to Salop.	4½	1: 2: 6
		[Total]	86: 7: 10

February 29th, 1872.	£: s: d Stone Bill 5: 11:5 Bawdley and Co. 1:2 [Sub Total] 5: 12:7 Balance 80: 15: 3 1872 April 6th, [Total] 86: 7: 10George Cank £100		
March 30th, 1872.	24"68, 0. Got and Loaded.	4/-	41: 2: 8
	19"70, 3. Haulage to Yorton.	1¾	14: 7: 8
	75yards Haycock Combing	2/6	5: 12: 6
	£: s:d	[Total]	61: 7: 5
	Brought Forward 19: 4:9		
	April 20th Cheque 42: 2:8 1872, April 20th		
	[Total] 61: 7:5 George Cank £42: 2: 8		
April 30th, 1872.	23"64, 1. Got and Loaded.	4/-	39: 8: 0
	16"64, 8. Haulage to Yorton.	1¾	12: 2: 8

	52, 1. [Ditto] Salop.	4½	19: 6
	65, 7. : Leaton	4½	1: 4: 4
	249, ½ yards Crushing.	1/6	18: 14: 3
		[Total]	72: 8: 9
	28:12:0 tons Burs	3/6	5: 0: 0
	£: s: d 1872 May 18th	[Total]	77: 8: 9
	May 4th Cheque 60: 0: 0George Cank £100		
	[Ditto] 18th : 40: 0: 0		
	[Total] 100: 0: 0		
	77: 8: 9		
	Cash Forward 22: 11: 3		
May 31st, 1872.	33"68, 0. Got and Loaded.	4/-	56: 2: 8
	29"12, 6. Haulage to Yorton.	1¾	21: 4: 8
	137, 3. [Ditto] Salop.	4½	2: 11: 4
	14, 10 : Black Birches.	1¾	2: 1

	54 yards crushing	1/-	2: 14: 0
	£: s: d	[Total]	82: 14: 9
	Brought Forward 22: 11: 3		
	Cheque 80: 0: 0		
	Stone Bill 4: 11: 8		
	[Total] 107:2: 11		
	82:14: 9June 1st, 1872		
	Balance 24: 8: 2George Cank £82: 14: 9		
June 30th, 1872.	19”69, .6. Got and Loaded.	4/ -	32: 16: 4
	16”84, 7. Haulage to Yorton.	1¾	12: 5: 7
	135, 10. [Ditto] Salop.	4½	2: 10: 9
	128, 2. : Trickhill.	3½	1: 17: 4
		[Total]	49: 10: 0
	27”17, 10. Got Stack.	3½	39: 12: 5: ½
	£: s: d	[Total]	89: 2:5:½

	Brought Forward 24: 8: 2		
June 15th	Cheque 30: 0:0		
	Stone Bill 13: 101872 June 30th		
July 12th	Cheque 50: 0: 0George Cank. £89:2:5: ½		
	[Total] 105:2: 0		
	89: 2: 5 ½		
	Cash Forward 15: 2:6 ½		
July 31st, 1872.	36"03, 10. Got and Loaded.	4/-	60: 1: 0
	32"67, 10. Haulage to Yorton.	1¾	23: 16: 5
	59, 10. [Ditto] Salop.	4½	1: 2: 2

	55, 5. :Trickhill.	3½	16: 0
	20:2:6 tons Crushing.	1/-	1: 0: 8
	£: s: d	[Total]	86: 16: 3
	Brought Forward 15: 19: 6 ½		
July 27th	Cheque 50: 0: 0		
	Stone Bill 2: 17: 7		
July 10th	Cheque 50: 0: 0		
	[Total] 118: 17: 1: ½		
	86: 16: 3		
	Cash Forward 32: 0: 10½		
August 31st, 1872.	27"83, 9. Got and Loaded.	4/-	46: 7: 8
	22"16, 3. Haulage to Yorton.	1¾	16: 3: 2
	314, 8. [Ditto]Salop.	4½	5: 17: 9
	60, 9. : Trickhill.	3½	17: 6
	41, 4. :Bridlegate.	1¾	6: 0

	33, 0. Omitted.	4/-	11: 0
	14:1:0 tons Tlay Burs.	2/6	2: 9: 0
	£: s: d	[Total]	72: 12: 1
	Brought Forward 32: 0: 10½		
August 31st, 1872.	Stone Bill. 2: 4: 5		
	…… over 3: 4		
	Cheque 50: 0:0		
	[Total] 84: 8: 7½		
	72: 12: 1		
	Cash Forward 11: 16:6½		
September 30th, 1872.	31" 80, 10. Got and Loaded.	4/-	53: 0: 0
	28" 46, 10. Haulage to Yorton.	1¾	20: 15: 0
	119: 5. [Ditto] Bridling Gate.	1¼	17: 4
	…….. Shedding Bill.		8: 15: 6
	£: s: d	[Total]	83: 7: 10

Sept 21st	Brought Forward 11: 16: 6½ Cheque 50: 0: 0 Stone Bill 2: 9: 2		
Sept 25th Oct 19th	Error 4: 6 Cheque 50: 0: 01872 Oct 19th [Total] 114: 10: 7½George Cank £83:7:10 83: 7: 10[Signed over stamp] Cash Forward 31: 2: 4½		
November 30th, 1872.	28" 68, 9. Got and Loaded.	4/-	47: 16: 0
	25" 26, 0. Haulage to Yorton.	1¾	18: 8: 4
	35, 1. [Ditto] Salop.	4½	13: 1½
	56, 8: Bridling Gate.		8: 2
	113 yards Crushing.	1/-	5: 13: 0
		[Total]	72: 18: 7½

	108, 6. Clay to Yorton.	1¾	15: 9½
	£: s: d	[Total]	73: 14: 4½
	Brought Forward 6: 16: 6½		
Nov 18th	Cheque 40: 0: 0 1872 Dec 4th		
Nov 30th	[Ditto] 40: 0: 0 George Cank £73:14: 4½		
	[Total] 86: 16: 6½		
	73: 14: 4½ Cash Forward 13: 2: 2		
December 30th, 1872.	30" 11, 4. Got and Loaded.	4/-	50: 0: 4
	29" 53, 10. Haulage to Yorton.	1¾	21: 10: 7
	45, 6. [Ditto] Salop.	4½	17: 0
	69 yards Crushing.		3: 9: 0
	£: s: d	[Total]	75: 16: 0
	Brought Forward 13: 2: 2		
	Cheque 40: 0: 0		
	[Ditto] 40: 0: 0 1873 Jan		

	Stone Bill 5: 8: 3 George Cank £75:16:11 Cottage Rent 2: 8: 3[Signed of stamp] Cheque 40: 0: 0 [Total] 140: 15: 5		
January 31st, 1873.	17" 88, 1. Got and Loaded. 16"68, 6. Haulage to Yorton. 1"19, 7.[Ditto] Salop. 132, 1/3yards Crushing. £: s: d	4/- 1¾ 4½ 1/- [Total]	29: 16: 0 12: 3: 3 2: 4: 8 6: 12: 4 50: 16: 3
	Brought Forward 64: 18: 6 Cheque 40: 0: 01873 February 22nd [Ditto] 50: 0: 0 George Cank £50:16: 3 [Total] 154: 18 : 6 [Signed over stamp] 50: 18: 3 [Total] 104: 2: 3		

February 28th, 1873.	21” 24, 4. Got and Loaded.	4/-	35: 8: 0
	26” 95, 11. Haulage to Yorton.	1¾	15: 9: 9
	£: s: d	[Total]	50: 17: 9
	Brought Forward 104: 2: 3 1873 Feb 28th		
March 8th	Cheque 50: 0: 0George Cank £50: 17: 9		
	[Total] 154: 2: 3 [Signed over stamp]		
	50: 17: 9		
	Cash Forward 103: 4: 6		
March 31st, 1873.	29” 48, 1. Got and Loaded.	4/-	49: 2: 8
	26” 95, 11. Haulage to Yorton.	1¾	19: 13: 8
	183, 1 [Ditto] Salop.	4½	3: 8: 7
		[Total]	72: 4: 3
	£: s: d		
	Brought Forward 103: 4: 6		
	Cheque 40: 0: 0		

March 22nd	Stone Bill	1: 5:	61873 April 5th		
	Cheque	50: 0:	0George Cank £72: 4: 3		
April 5th	[Total]	194: 10: 0			
		72: 4:	3		
	Cash Forward	122: 5:	9		

(Source: Shropshire Archives, The Account Book of Samuel and George Cank's business for 1870 to 1873 MS No. 4756/1/22, accessed 12/10/2011).

Appendix 2: Photographs of former Cranes and sites

Figure 1: Grinshill Stone Quarries, former crane foundations. Taken by author; 10/08/2011.

Figure 2: Grinshill Stone Quarries, former crane foundations. Taken by author: 10/08/2011

Appendix 3: Photographs of track used to transport stone by horses

Figure 3: Corbett Wood track used by horses to transport stone. Taken by author: 10/08/2011

Appendix 4: Photographs of stone blocks extracted by Bristar Powder

Figure 4: Holes were Bristar Powder s were drilled and placed. Taken by author 10/08/2011.

Figure 5: Block of stone extracted by use of Bristar Powder. Taken by author 10/08/2011.

Figure 6: Showing stone site after stone block removed. Taken by author 10/08/2011.